Welding

Welding

Edited by
Gilbert Mead

Larsen & Keller
www.larsen-keller.com

Welding
Edited by Gilbert Mead
ISBN: 978-1-63549-295-8 (Hardback)

© 2017 Larsen & Keller

▤ Larsen & Keller

Published by Larsen and Keller Education,
5 Penn Plaza,
19th Floor,
New York, NY 10001, USA

Cataloging-in-Publication Data

Welding / edited by Gilbert Mead.
     p. cm.
Includes bibliographical references and index.
ISBN 978-1-63549-295-8
1. Welding. 2. Plastics--Welding. 3. Brazing.
4. Solder and soldering. I. Mead, Gilbert.
TS227 .W45 2017
671.52--dc23

The publisher's policy is to use permanent paper from mills that operate a sustainable forestry policy. Furthermore, the publisher ensures that the text paper and cover boards used have met acceptable environmental accreditation standards.

Printed and bound in the United States of America.

For more information regarding Larsen and Keller Education and its products, please visit the publisher's website www.larsen-keller.com

Table of Contents

Preface

Welding refers to the science and practice of joining metals and thermoplastics by using fusion and related techniques like soldering and brazing. The most common methods associated with this area are gas metal arc welding, shielded metal arc welding, electroslag welding, gas tungsten arc welding, etc. The aim of this text is to give students a deep insight about the various methods and technologies used in this field. Most of the topics introduced in it cover new techniques and applications of welding. Different approaches, evaluations and methodologies on the subject have been included in it. The textbook will serve as a valuable source of reference for those interested in this field.

Given below is the chapter wise description of the book:

Chapter 1- Welding is the process of joining materials together with the help of techniques such as brazing and soldering. The form of welding that uses electromagnetic induction is known as induction welding. This is an introductory chapter which will introduce briefly all the significant aspects of welding.

Chapter 2- Shielded metal arc welding is a manual method of welding which uses electrodes covered with a flux to lay the weld. The alternative methods of welding are gas tungsten arc welding, flux-cored arc welding, submerged arc welding, spot welding, forge welding and robot welding. This section discusses the methods of welding in a critical manner providing key analysis to the subject matter.

Chapter 3- The process of binding metal to metal by using electricity is known as arc welding. The other processes of welding are electric resistance welding, filet weld, plasma cutting, fusion welding, explosion welding, cold welding etc. The aspects elucidated in this chapter are of vital importance, and provide a better understanding of welding.

Chapter 4- A welding power supply is a device and this device helps in performing welding. It is a very simple device, it is as simple as a car battery. Along with welding power supply, shot welding, welding procedure specification, weld purging and orbital welding are the essential techniques of welding. The section strategically encompasses and incorporates the major techniques related to welding, providing a complete understanding.

Chapter 5- Plastic welding as a process has three stages, namely surface preparation, application of heat and pressure and cooling. Along with plastic welding, friction welding, heat fusion, spin welding and ultrasonic welding have been elucidated in the following text. This section will provide an integrated understanding of plastic welding.

Chapter 6- Brazing is done to join two metals; metal is melted and filled into the joint that connects the two metals. There is a difference between welding and brazing, as welding does not involve melting the work pieces. This chapter also explains to the reader the concepts of soldering, dip soldering, ultrasonic soldering and wave soldering.

Chapter 7- Welding is the art of joining two or more metals, this requires using numerous metals during the process. Some of these metals are aluminum, titanium, beryllium and zirconium. This section closely examines all the metals used in welding and provides an easy understanding of the subject matter.

At the end, I would like to thank all those who dedicated their time and efforts for the successful completion of this book. I also wish to convey my gratitude towards my friends and family who supported me at every step.

Editor

Introduction to Welding

Welding is the process of joining materials together with the help of techniques such as brazing and soldering. The form of welding that uses electromagnetic induction is known as induction welding. This is an introductory chapter which will introduce briefly all the significant aspects of welding.

Welding

Gas metal arc welding (MIG welding)

Welding is a fabrication or sculptural process that joins materials, usually metals or thermoplastics, by causing fusion, which is distinct from lower temperature metal-joining techniques such as brazing and soldering, which do not melt the base metal. In addition to melting the base metal, a filler material is often added to the joint to form a pool of molten material (the weld pool) that cools to form a joint that can be as strong, or even stronger, than the base material. Pressure may also be used in conjunction with heat, or by itself, to produce a weld.

Although less common, there are also solid state welding processes such as friction welding or shielded active gas welding in which metal does not melt.

Some of the best known welding methods include:

- Shielded metal arc welding (SMAW) – also known as "stick welding or electric welding", uses an electrode that has flux around it to protect the weld puddle. The electrode holder holds the electrode as it slowly melts away. Slag protects the weld puddle from atmospheric contamination.

- Gas tungsten arc welding (GTAW) – also known as TIG (tungsten, inert gas), uses a non-consumable tungsten electrode to produce the weld. The weld area is protected from atmospheric contamination by an inert shielding gas such as argon or helium.

- Gas metal arc welding (GMAW) – commonly termed MIG (metal, inert gas), uses a wire feeding gun that feeds wire at an adjustable speed and flows an argon-based shielding gas or a mix of argon and carbon dioxide (CO_2) over the weld puddle to protect it from atmospheric contamination.

- Flux-cored arc welding (FCAW) – almost identical to MIG welding except it uses a special tubular wire filled with flux; it can be used with or without shielding gas, depending on the filler.

- Submerged arc welding (SAW) – uses an automatically fed consumable electrode and a blanket of granular fusible flux. The molten weld and the arc zone are protected from atmospheric contamination by being "submerged" under the flux blanket.

- Electroslag welding (ESW) – a highly productive, single pass welding process for thicker materials between 1 inch (25 mm) and 12 inches (300 mm) in a vertical or close to vertical position.

Many different energy sources can be used for welding, including a gas flame, an electric arc, a laser, an electron beam, friction, and ultrasound. While often an industrial process, welding may be performed in many different environments, including in open air, under water, and in outer space. Welding is a hazardous undertaking and precautions are required to avoid burns, electric shock, vision damage, inhalation of poisonous gases and fumes, and exposure to intense ultraviolet radiation.

Until the end of the 19th century, the only welding process was forge welding, which blacksmiths had used for centuries to join iron and steel by heating and hammering. Arc welding and oxyfuel welding were among the first processes to develop late in the century, and electric resistance welding followed soon after. Welding technology advanced quickly during the early 20th century as the world wars drove the demand for reliable and inexpensive joining methods. Following the wars, several modern welding techniques were developed, including manual methods like SMAW, now one of the most popular welding methods, as well as semi-automatic and automatic processes such as GMAW, SAW, FCAW and ESW. Developments continued with the invention of laser beam welding, electron beam welding, magnetic pulse welding (MPW), and friction stir welding in the latter half of the century. Today, the science continues to advance. Robot welding is commonplace in industrial settings, and researchers continue to develop new welding methods and gain greater understanding of weld quality.

History

The history of joining metals goes back several millennia. Called forge welding, the earliest examples come from the Bronze and Iron Ages in Europe and the Middle East. The ancient Greek historian Herodotus states in *The Histories* of the 5th century BC that Glaucus of Chios "was the man who single-handedly invented iron welding". Welding was used in the construction of the Iron pillar of Delhi, erected in Delhi, India about 310 AD and weighing 5.4 metric tons.

The iron pillar of Delhi

The Middle Ages brought advances in forge welding, in which blacksmiths pounded heated metal repeatedly until bonding occurred. In 1540, Vannoccio Biringuccio published *De la pirotechnia*, which includes descriptions of the forging operation. Renaissance craftsmen were skilled in the process, and the industry continued to grow during the following centuries.

In 1800, Sir Humphry Davy discovered the short-pulse electrical arc and presented his results in 1801. In 1802, Russian scientist Vasily Petrov created the continuous electric arc, and subsequently published "News of Galvanic-Voltaic Experiments" in 1803, in which he described experiments carried out in 1802. Of great importance in this work was the description of a stable arc discharge and the indication of its possible use for many applications, one being melting metals. In 1808, Davy, who was unaware of Petrov's work, rediscovered the continuous electric arc. In 1881–82 inventors Nikolai Benardos (Russian) and Stanisław Olszewski (Polish) created the first electric arc welding method known as carbon arc welding using carbon electrodes. The advances in arc welding continued with the invention of metal electrodes in the late 1800s by a Russian, Nikolai Slavyanov (1888), and an American, C. L. Coffin (1890). Around 1900, A. P. Strohmenger released a coated metal electrode in Britain, which gave a more stable arc. In 1905, Russian scientist Vladimir Mitkevich proposed using a three-phase electric arc for welding. In 1919, alternating current welding was invented by C. J. Holslag but did not become popular for another decade.

Resistance welding was also developed during the final decades of the 19th century, with the first patents going to Elihu Thomson in 1885, who produced further advances over the next 15 years. Thermite welding was invented in 1893, and around that time another process, oxyfuel welding, became well established. Acetylene was discovered in 1836 by Edmund Davy, but its use was not practical in welding until about 1900, when a suitable torch was developed. At first, oxyfuel welding was one of the more popular welding methods due to its portability and relatively low cost. As the 20th century progressed, however, it fell out of favor for industrial applications. It was largely replaced with arc welding, as metal coverings (known as flux) for the electrode that stabilize the arc and shield the base material from impurities continued to be developed.

Bridge of Maurzyce

World War I caused a major surge in the use of welding processes, with the various military powers attempting to determine which of the several new welding processes would be best. The British primarily used arc welding, even constructing a ship, the "Fullagar" with an entirely welded hull. Arc welding was first applied to aircraft during the war as well, as some German airplane fuselages were constructed using the process. Also noteworthy is the first welded road bridge in the world, the Maurzyce Bridge designed by Stefan Bryła of the Lwów University of Technology in 1927, and built across the river Słudwia near Łowicz, Poland in 1928.

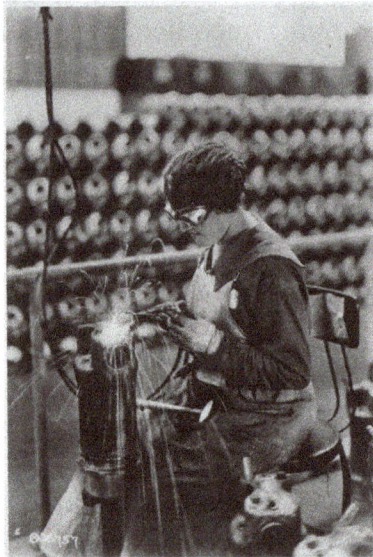

Acetylene welding on cylinder water jacket, 1918

During the 1920s, major advances were made in welding technology, including the introduction of automatic welding in 1920, in which electrode wire was fed continuously. Shielding gas became a subject receiving much attention, as scientists attempted to protect welds from the effects of oxygen and nitrogen in the atmosphere. Porosity and brittleness were the primary problems, and the solutions that developed included the use of hydrogen, argon, and helium as welding atmospheres. During the following decade, further advances allowed for the welding of reactive metals like aluminum and magnesium. This in conjunction with developments in automatic welding, alternating current, and fluxes fed a major expansion of arc welding during the 1930s and then during World War II. In 1930, the first all-welded merchant vessel, M/S Carolinian, was launched.

During the middle of the century, many new welding methods were invented. In 1930, Kyle Taylor was responsible for the release of stud welding, which soon became popular in shipbuilding and construction. Submerged arc welding was invented the same year and continues to be popular today. In 1932 a Russian, Konstantin Khrenov successfully implemented the first underwater electric arc welding. Gas tungsten arc welding, after decades of development, was finally perfected in 1941, and gas metal arc welding followed in 1948, allowing for fast welding of non-ferrous materials but requiring expensive shielding gases. Shielded metal arc welding was developed during the 1950s, using a flux-coated consumable electrode, and it quickly became the most popular metal arc welding process. In 1957, the flux-cored arc welding process debuted, in which the self-shielded wire electrode could be used with automatic equipment, resulting in greatly increased welding speeds, and that same year, plasma arc welding was invented. Electroslag welding was introduced in 1958, and it was followed by its cousin, electrogas welding, in 1961. In 1953 the Soviet scientist N. F. Kazakov proposed the diffusion bonding method.

Other recent developments in welding include the 1958 breakthrough of electron beam welding, making deep and narrow welding possible through the concentrated heat source. Following the invention of the laser in 1960, laser beam welding debuted several decades later, and has proved to be especially useful in high-speed, automated welding. Magnetic pulse welding (MPW) is industrially used since 1967. Friction stir welding was invented in 1991 by Wayne Thomas at The Welding Institute (TWI, UK) and found high-quality applications all over the world. All of these four new processes continue to be quite expensive due the high cost of the necessary equipment, and this has limited their applications.

Processes

Arc

Man welding a metal structure in a newly constructed house in Bengaluru, India

These processes use a welding power supply to create and maintain an electric arc between an electrode and the base material to melt metals at the welding point. They can use either direct (DC) or alternating (AC) current, and consumable or non-consumable electrodes. The welding region is sometimes protected by some type of inert or semi-inert gas, known as a shielding gas, and filler material is sometimes used as well.

Power Supplies

To supply the electrical power necessary for arc welding processes, a variety of different power supplies can be used. The most common welding power supplies are constant current power supplies and constant voltage power supplies. In arc welding, the length of the arc is directly related to the voltage, and the amount of heat input is related to the current. Constant current power supplies are most often used for manual welding processes such as gas tungsten arc welding and shielded metal arc welding, because they maintain a relatively constant current even as the voltage varies. This is important because in manual welding, it can be difficult to hold the electrode perfectly steady, and as a result, the arc length and thus voltage tend to fluctuate. Constant voltage power supplies hold the voltage constant and vary the current, and as a result, are most often used for automated welding processes such as gas metal arc welding, flux cored arc welding, and submerged arc welding. In these processes, arc length is kept constant, since any fluctuation in the distance between the wire and the base material is quickly rectified by a large change in current. For example, if the wire and the base material get too close, the current will rapidly increase, which in turn causes the heat to increase and the tip of the wire to melt, returning it to its original separation distance.

The type of current used plays an important role in arc welding. Consumable electrode processes such as shielded metal arc welding and gas metal arc welding generally use direct current, but the electrode can be charged either positively or negatively. In welding, the positively charged anode will have a greater heat concentration, and as a result, changing the polarity of the electrode affects weld properties. If the electrode is positively charged, the base metal will be hotter, increasing weld penetration and welding speed. Alternatively, a negatively charged electrode results in more shallow welds. Nonconsumable electrode processes, such as gas tungsten arc welding, can use either type of direct current, as well as alternating current. However, with direct current, because the electrode only creates the arc and does not provide filler material, a positively charged electrode causes shallow welds, while a negatively charged electrode makes deeper welds. Alternating current rapidly moves between these two, resulting in medium-penetration welds. One disadvantage of AC, the fact that the arc must be re-ignited after every zero crossing, has been addressed with the invention of special power units that produce a square wave pattern instead of the normal sine wave, making rapid zero crossings possible and minimizing the effects of the problem.

Processes

One of the most common types of arc welding is shielded metal arc welding (SMAW); it is also known as manual metal arc welding (MMA) or stick welding. Electric current is used to strike an arc between the base material and consumable electrode rod, which is made of filler material (typically steel) and is covered with a flux that protects the weld area from oxidation and contamination by producing carbon dioxide (CO_2) gas during the welding process. The electrode core itself acts as filler material, making a separate filler unnecessary.

Shielded metal arc welding

The process is versatile and can be performed with relatively inexpensive equipment, making it well suited to shop jobs and field work. An operator can become reasonably proficient with a modest amount of training and can achieve mastery with experience. Weld times are rather slow, since the consumable electrodes must be frequently replaced and because slag, the residue from the flux, must be chipped away after welding. Furthermore, the process is generally limited to welding ferrous materials, though special electrodes have made possible the welding of cast iron, nickel, aluminum, copper, and other metals.

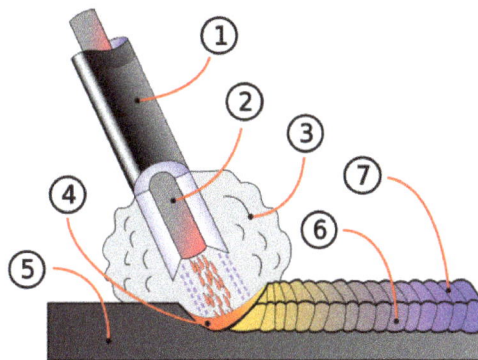

Diagram of arc and weld area, in shielded metal arc welding.

1. Coating Flow
2. Rod
3. Shield Gas
4. Fusion
5. Base metal
6. Weld metal
7. Solidified Slag

Gas metal arc welding (GMAW), also known as metal inert gas or MIG welding, is a semi-automatic or automatic process that uses a continuous wire feed as an electrode and an inert or semi-inert gas mixture to protect the weld from contamination. Since the electrode is continuous, welding speeds are greater for GMAW than for SMAW.

A related process, flux-cored arc welding (FCAW), uses similar equipment but uses wire consisting of a steel electrode surrounding a powder fill material. This cored wire is more expensive than the standard solid wire and can generate fumes and/or slag, but it permits even higher welding speed and greater metal penetration.

Gas tungsten arc welding (GTAW), or tungsten inert gas (TIG) welding, is a manual welding process that uses a nonconsumable tungsten electrode, an inert or semi-inert gas mixture, and a separate filler material. Especially useful for welding thin materials, this method is characterized by a stable arc and high quality welds, but it requires significant operator skill and can only be accomplished at relatively low speeds.

GTAW can be used on nearly all weldable metals, though it is most often applied to stainless steel and light metals. It is often used when quality welds are extremely important, such as in bicycle, aircraft and naval applications. A related process, plasma arc welding, also uses a tungsten electrode but uses plasma gas to make the arc. The arc is more concentrated than the GTAW arc, making transverse control more critical and thus generally restricting the technique to a mechanized process. Because of its stable current, the method can be used on a wider range of material thicknesses than can the GTAW process and it is much faster. It can be applied to all of the same materials as GTAW except magnesium, and automated welding of stainless steel is one important application of the process. A variation of the process is plasma cutting, an efficient steel cutting process.

Submerged arc welding (SAW) is a high-productivity welding method in which the arc is struck beneath a covering layer of flux. This increases arc quality, since contaminants in the atmosphere are blocked by the flux. The slag that forms on the weld generally comes off by itself, and combined with the use of a continuous wire feed, the weld deposition rate is high. Working conditions are much improved over other arc welding processes, since the flux hides the arc and almost no smoke is produced. The process is commonly used in industry, especially for large products and in the manufacture of welded pressure vessels. Other arc welding processes include atomic hydrogen welding, electroslag welding, electrogas welding, and stud arc welding.

Gas Welding

The most common gas welding process is oxyfuel welding, also known as oxyacetylene welding. It is one of the oldest and most versatile welding processes, but in recent years it has become less popular in industrial applications. It is still widely used for welding pipes and tubes, as well as repair work.

The equipment is relatively inexpensive and simple, generally employing the combustion of acetylene in oxygen to produce a welding flame temperature of about 3100 °C. The flame, since it is less concentrated than an electric arc, causes slower weld cooling, which can lead to greater residual stresses and weld distortion, though it eases the welding of high alloy steels. A similar process, generally called oxyfuel cutting, is used to cut metals.

Resistance

Resistance welding involves the generation of heat by passing current through the resistance caused by the contact between two or more metal surfaces. Small pools of molten metal are formed at the weld area as high current (1000–100,000 A) is passed through the metal. In general, resistance welding methods are efficient and cause little pollution, but their applications are somewhat limited and the equipment cost can be high.

Spot welder

Spot welding is a popular resistance welding method used to join overlapping metal sheets of up to 3 mm thick. Two electrodes are simultaneously used to clamp the metal sheets together and to pass current through the sheets. The advantages of the method include efficient energy use, limited workpiece deformation, high production rates, easy automation, and no required filler materials. Weld strength is significantly lower than with other welding methods, making the process suitable for only certain applications. It is used extensively in the automotive industry—ordinary cars can have several thousand spot welds made by industrial robots. A specialized process, called shot welding, can be used to spot weld stainless steel.

Like spot welding, seam welding relies on two electrodes to apply pressure and current to join metal sheets. However, instead of pointed electrodes, wheel-shaped electrodes roll along and often feed the workpiece, making it possible to make long continuous welds. In the past, this process was used in the manufacture of beverage cans, but now its uses are more limited. Other resistance welding methods include butt welding, flash welding, projection welding, and upset welding.

Energy Beam

Energy beam welding methods, namely laser beam welding and electron beam welding, are relatively new processes that have become quite popular in high production applications. The two processes are quite similar, differing most notably in their source of power. Laser beam welding employs a highly focused laser beam, while electron beam welding is done in a vacuum and uses an electron beam. Both have a very high energy density, making deep weld penetration possible and minimizing the size of the weld area. Both processes are extremely fast, and are easily automated, making them highly productive. The primary disadvantages are their very high equipment costs (though these are decreasing) and a susceptibility to thermal cracking. Developments in this area include laser-hybrid welding, which uses principles from both laser beam welding and arc welding for even better weld properties, laser cladding, and x-ray welding.

Solid-state

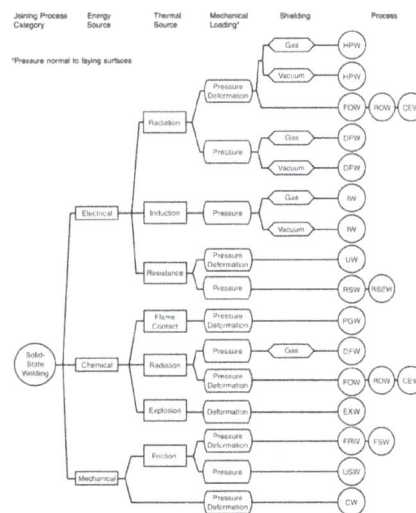

Solid-state welding processes classification chart

Like the first welding process, forge welding, some modern welding methods do not involve the melting of the materials being joined. One of the most popular, ultrasonic welding, is used to connect thin sheets or wires made of metal or thermoplastic by vibrating them at high frequency and under high pressure. The equipment and methods involved are similar to that of resistance welding, but instead of electric current, vibration provides energy input. Welding metals with this process does not involve melting the materials; instead, the weld is formed by introducing mechanical vibrations horizontally under pressure. When welding plastics, the materials should have similar melting temperatures, and the vibrations are introduced vertically. Ultrasonic welding is commonly used for making electrical connections out of aluminum or copper, and it is also a very common polymer welding process.

Another common process, explosion welding, involves the joining of materials by pushing them together under extremely high pressure. The energy from the impact plasticizes the materials, forming
a weld, even though only a limited amount of heat is generated. The process is commonly used for
welding dissimilar materials, such as the welding of aluminum with steel in ship hulls or compound
plates. Other solid-state welding processes include friction welding (including friction stir welding),
magnetic pulse welding, co-extrusion welding, cold welding, diffusion bonding, exothermic welding,
high frequency welding, hot pressure welding, induction welding, and roll welding.

Geometry

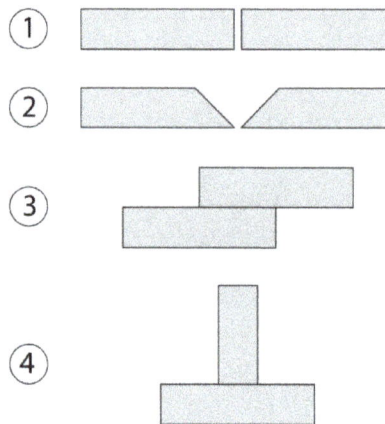

Common welding joint types – (1) Square butt joint, (2) V butt joint, (3) Lap joint, (4) T-joint

Welds can be geometrically prepared in many different ways. The five basic types of weld joints are
the butt joint, lap joint, corner joint, edge joint, and T-joint (a variant of this last is the cruciform
joint). Other variations exist as well—for example, double-V preparation joints are characterized
by the two pieces of material each tapering to a single center point at one-half their height. Sin-
gle-U and double-U preparation joints are also fairly common—instead of having straight edges
like the single-V and double-V preparation joints, they are curved, forming the shape of a U. Lap
joints are also commonly more than two pieces thick—depending on the process used and the
thickness of the material, many pieces can be welded together in a lap joint geometry.

Many welding processes require the use of a particular joint design; for example, resistance spot
welding, laser beam welding, and electron beam welding are most frequently performed on lap
joints. Other welding methods, like shielded metal arc welding, are extremely versatile and can
weld virtually any type of joint. Some processes can also be used to make multipass welds, in which
one weld is allowed to cool, and then another weld is performed on top of it. This allows for the
welding of thick sections arranged in a single-V preparation joint, for example.

The cross-section of a welded butt joint, with the darkest gray representing the weld or fusion zone, the
medium gray the heat-affected zone, and the lightest gray the base material.

After welding, a number of distinct regions can be identified in the weld area. The weld itself is called the fusion zone—more specifically, it is where the filler metal was laid during the welding process. The properties of the fusion zone depend primarily on the filler metal used, and its compatibility with the base materials. It is surrounded by the heat-affected zone, the area that had its microstructure and properties altered by the weld. These properties depend on the base material's behavior when subjected to heat. The metal in this area is often weaker than both the base material and the fusion zone, and is also where residual stresses are found.

Quality

The blue area results from oxidation at a corresponding temperature of 600 °F (316 °C). This is an accurate way to identify temperature, but does not represent the HAZ width. The HAZ is the narrow area that immediately surrounds the welded base metal.

Many distinct factors influence the strength of welds and the material around them, including the welding method, the amount and concentration of energy input, the weldability of the base material, filler material, and flux material, the design of the joint, and the interactions between all these factors. To test the quality of a weld, either destructive or nondestructive testing methods are commonly used to verify that welds are free of defects, have acceptable levels of residual stresses and distortion, and have acceptable heat-affected zone (HAZ) properties. Types of welding defects include cracks, distortion, gas inclusions (porosity), non-metallic inclusions, lack of fusion, incomplete penetration, lamellar tearing, and undercutting.

The metalworking industry has instituted specifications and codes to guide welders, weld inspectors, engineers, managers, and property owners in proper welding technique, design of welds, how to judge the quality of Welding Procedure Specification, how to judge the skill of the person performing the weld, and how to ensure the quality of a welding job. Methods such as visual inspection, radiography, ultrasonic testing, phased-array ultrasonics, dye penetrant inspection, magnetic particle inspection, or industrial computed tomography can help with detection and analysis of certain defects.

Heat-affected Zone

The effects of welding on the material surrounding the weld can be detrimental—depending on the materials used and the heat input of the welding process used, the HAZ can be of varying size and strength. The thermal diffusivity of the base material plays a large role—if the diffusivity is high, the material cooling rate is high and the HAZ is relatively small. Conversely, a low diffusivity leads to slower cooling and a larger HAZ. The amount of heat injected by the welding process plays an important role as well, as processes like oxyacetylene welding have an unconcentrated heat input and increase the size of the HAZ. Processes like laser beam welding give a highly concentrated, limited amount of heat, resulting in a small HAZ. Arc welding falls between these two extremes, with the individual processes varying somewhat in heat input.

The efficiency is dependent on the welding process used, with shielded metal arc welding having a value of 0.75, gas metal arc welding and submerged arc welding, 0.9, and gas tungsten arc welding, 0.8.

Lifetime Extension With Aftertreatment Methods

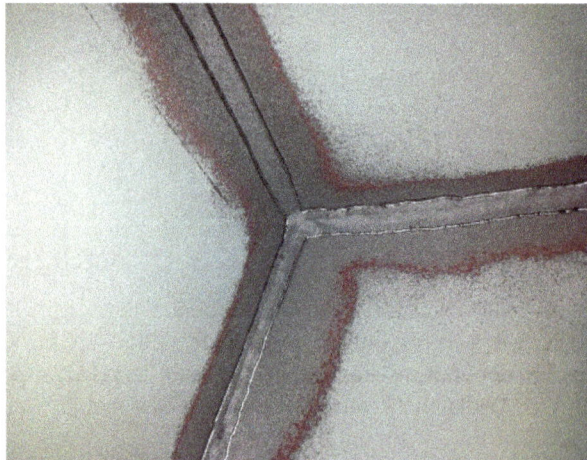

Example: High Frequency Impact Treatment for lifetime extension

The durability and life of dynamically loaded, welded steel structures is determined in many cases by the welds, particular the weld transitions. Through selective treatment of the transitions by grinding (abrasive cutting), shot peening, High Frequency Impact Treatment, etc. the durability of many designs increase significantly.

Metallurgy

Most solids used are engineering materials consisting of crystalline solids in which the atoms or ions are arranged in a repetitive geometric pattern which is known as a lattice structure. The only exception is material that is made from glass which is a combination of a supercooled liquid and polymers which are aggregates of large organic molecules.

Crystalline solids cohesion is obtained by a metallic or chemical bond which is formed between the constituent atoms. Chemical bonds can be grouped into two types consisting of ionic and covalent. To form an ionic bond, either a valence or bonding electron separates from one atom and becomes attached to another atom to form oppositely charged ions. The bonding in the static position is when the ions occupy an equilibrium position where the resulting force between them is zero. When the ions are exerted in tension force, the inter-ionic spacing increases creating an electrostatic attractive force, while a repulsing force under compressive force between the atomic nuclei is dominant.

Covalent bonding takes place when one of the constituent atoms loses one or more electrons, with the other atom gaining the electrons, resulting in an electron cloud that is shared by the molecule as a whole. In both ionic and covalent bonding the location of the ions and electrons are constrained relative to each other, thereby resulting in the bond being characteristically brittle.

Metallic bonding can be classified as a type of covalent bonding for which the constituent atoms of the same type and do not combine with one another to form a chemical bond. Atoms will lose an electron(s) forming an array of positive ions. These electrons are shared by the lattice which makes the electron cluster mobile, as the electrons are free to move as well as the ions. For this, it gives metals their relatively high thermal and electrical conductivity as well as being characteristically ductile.

Three of the most commonly used crystal lattice structures in metals are the body-centred cubic, face-centred cubic and close-packed hexagonal. Ferritic steel has a body-centred cubic structure and austenitic steel, non-ferrous metals like aluminum, copper and nickel have the face-centred cubic structure.

Ductility is an important factor in ensuring the integrity of structures by enabling them to sustain local stress concentrations without fracture. In addition, structures are required to be of an acceptable strength, which is related to a material's yield strength. In general, as the yield strength of a material increases, there is a corresponding reduction in fracture toughness.

A reduction in fracture toughness may also be attributed to the embrittlement effect of impurities, or for body-centred cubic metals, from a reduction in temperature. Metals and in particular steels have a transitional temperature range where above this range the metal has acceptable notch-ductility while below this range the material becomes brittle. Within the range, the materials behavior is unpredictable. The reduction in fracture toughness is accompanied by a change in the fracture appearance. When above the transition, the fracture is primarily due to micro-void coalescence, which results in the fracture appearing fibrous. When the temperatures falls the fracture will show signs of cleavage facets. These two appearances are visible by the naked eye. Brittle fracture in steel plates may appear as chevron markings under the microscope. These arrow-like ridges on the crack surface point towards the origin of the fracture.

Fracture toughness is measured using a notched and pre-cracked rectangular specimen, of which the dimensions are specified in standards, for example ASTM E23. There are other means of estimating or measuring fracture toughness by the following: The Charpy impact test per ASTM A370; The crack-tip opening displacement (CTOD) test per BS 7448-1; The J integral test per ASTM E1820; The Pellini drop-weight test per ASTM E208.

Unusual Conditions

Underwater welding

While many welding applications are done in controlled environments such as factories and repair shops, some welding processes are commonly used in a wide variety of conditions, such as open air, underwater, and vacuums (such as space). In open-air applications, such as construction and outdoors repair, shielded metal arc welding is the most common process. Processes that employ inert gases to protect the weld cannot be readily used in such situations, because unpredictable atmospheric movements can result in a faulty weld. Shielded metal arc welding is also often used in underwater welding in the construction and repair of ships, offshore platforms, and pipelines, but others, such as flux cored arc welding and gas tungsten arc welding, are also common. Welding in space is also possible—it was first attempted in 1969 by Russian cosmonauts, when they performed experiments to test shielded metal arc welding, plasma arc welding, and electron beam welding in a depressurized environment. Further testing of these methods was done in the following decades, and today researchers continue to develop methods for using other welding processes in space, such as laser beam welding, resistance welding, and friction welding. Advances in these areas may be useful for future endeavours similar to the construction of the International Space Station, which could rely on welding for joining in space the parts that were manufactured on Earth.

Safety Issues

Welding can be dangerous and unhealthy if the proper precautions are not taken. However, using new technology and proper protection greatly reduces risks of injury and death associated with welding. Since many common welding procedures involve an open electric arc or flame, the risk of burns and fire is significant; this is why it is classified as a hot work process. To prevent injury, welders wear personal protective equipment in the form of heavy leather gloves and protective long-sleeve jackets to avoid exposure to extreme heat and flames. Additionally, the brightness of the weld area leads to a condition called arc eye or flash burns in which ultraviolet light causes inflammation of the cornea and can burn the retinas of the eyes. Goggles and welding helmets with dark UV-filtering face plates are worn to prevent this exposure. Since the 2000s, some helmets have included a face plate which instantly darkens upon exposure to the intense UV light. To protect bystanders, the welding area is often surrounded with translucent welding curtains. These curtains, made of a polyvinyl chloride plastic film, shield people outside the welding area from the UV light of the electric arc, but can not replace the filter glass used in helmets.

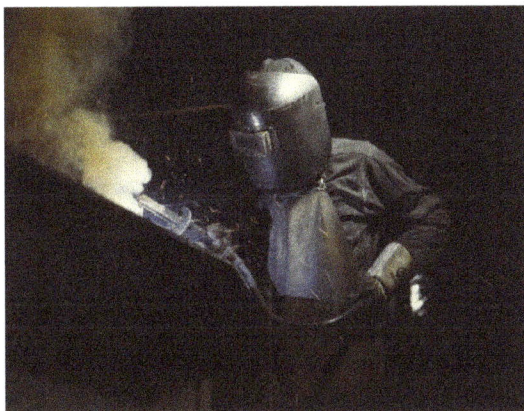

Arc welding with a welding helmet, gloves, and other protective clothing

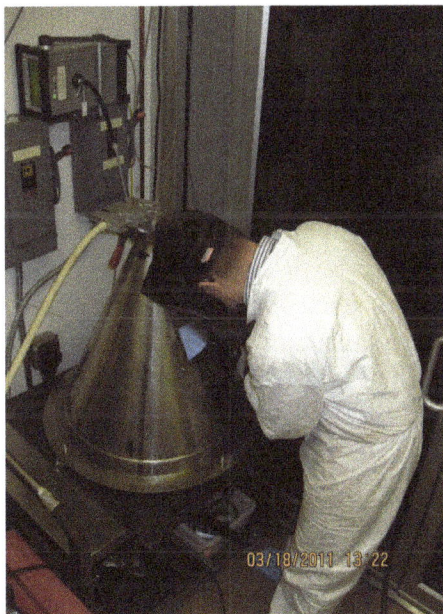

A chamber designed to contain welding fumes for analysis

Welders are often exposed to dangerous gases and particulate matter. Processes like flux-cored arc welding and shielded metal arc welding produce smoke containing particles of various types of oxides. The size of the particles in question tends to influence the toxicity of the fumes, with smaller particles presenting a greater danger. This is because smaller particles have the ability to cross the blood brain barrier. Fumes and gases, such as carbon dioxide, ozone, and fumes containing heavy metals, can be dangerous to welders lacking proper ventilation and training. Exposure to manganese welding fumes, for example, even at low levels (<0.2 mg/m^3), may lead to neurological problems or to damage to the lungs, liver, kidneys, or central nervous system. Nano particles can become trapped in the alveolar macrophages of the lungs and induce pulmonary fibrosis. The use of compressed gases and flames in many welding processes poses an explosion and fire risk. Some common precautions include limiting the amount of oxygen in the air, and keeping combustible materials away from the workplace.

Costs and Trends

As an industrial process, the cost of welding plays a crucial role in manufacturing decisions. Many different variables affect the total cost, including equipment cost, labor cost, material cost, and energy cost. Depending on the process, equipment cost can vary, from inexpensive for methods like shielded metal arc welding and oxyfuel welding, to extremely expensive for methods like laser beam welding and electron beam welding. Because of their high cost, they are only used in high production operations. Similarly, because automation and robots increase equipment costs, they are only implemented when high production is necessary. Labor cost depends on the deposition rate (the rate of welding), the hourly wage, and the total operation time, including time spent fitting, welding, and handling the part. The cost of materials includes the cost of the base and filler material, and the cost of shielding gases. Finally, energy cost depends on arc time and welding power demand.

For manual welding methods, labor costs generally make up the vast majority of the total cost. As a result, many cost-saving measures are focused on minimizing operation time. To do this, welding procedures with high deposition rates can be selected, and weld parameters can be fine-tuned to increase welding speed. Mechanization and automation are often implemented to reduce labor costs, but this frequently increases the cost of equipment and creates additional setup time. Material costs tend to increase when special properties are necessary, and energy costs normally do not amount to more than several percent of the total welding cost.

In recent years, in order to minimize labor costs in high production manufacturing, industrial welding has become increasingly more automated, most notably with the use of robots in resistance spot welding (especially in the automotive industry) and in arc welding. In robot welding, mechanized devices both hold the material and perform the weld and at first, spot welding was its most common application, but robotic arc welding increases in popularity as technology advances. Other key areas of research and development include the welding of dissimilar materials (such as steel and aluminum, for example) and new welding processes, such as friction stir, magnetic pulse, conductive heat seam, and laser-hybrid welding. Furthermore, progress is desired in making more specialized methods like laser beam welding practical for more applications, such as in the aerospace and automotive industries. Researchers also hope to better understand the often unpredictable properties of welds, especially microstructure, residual stresses, and a weld's tendency to crack or deform.

The trend of accelerating the speed at which welds are performed in the steel erection industry comes at a risk to the integrity of the connection. Without proper fusion to the base materials provided by sufficient arc time on the weld, a project inspector cannot ensure the effective diameter of the puddle weld therefore he or she cannot guarantee the published load capacities unless they witness the actual installation. This method of puddle welding is common in the United States and Canada for attaching steel sheets to bar joist and structural steel members. Regional agencies are responsible for ensuring the proper installation of puddle welding on steel construction sites. Currently there is no standard or weld procedure which can ensure the published holding capacity of any unwitnessed connection, but this is under review by the American Welding Society.

Glass and Plastic Welding

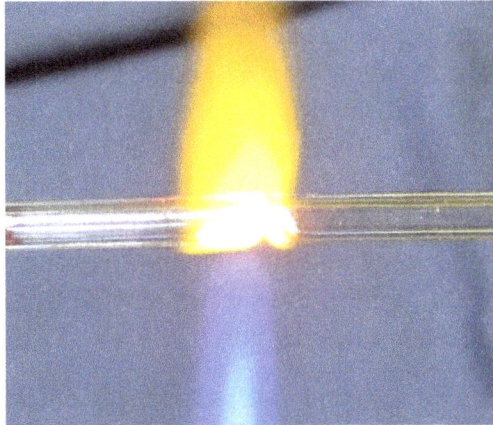

The welding together of two tubes made from lead glass

A bowl made from cast-glass. The two halves are joined together by the weld seam, running down the middle.

Glasses and certain types of plastics are commonly welded materials. Unlike metals, which have a specific melting point, glasses and plastics have a melting range, called the glass transition. When heating the solid material into this range, it will generally become softer and more pliable. When it crosses through the glass transition, it will become a very thick, sluggish, viscous liquid. Typically, this viscous liquid will have very little surface tension, becoming a sticky, honey-like consistency, so welding can usually take place by simply pressing two melted surfaces together. The two liquids will generally mix and join at first contact. Upon cooling through the glass transition, the welded piece will solidify as one solid piece of amorphous material.

Glass Welding

Glass welding is a common practice during glassblowing. It is used very often in the construction of lighting, neon signs, flashtubes, scientific equipment, and the manufacture of dishes and other glassware. It is also used during glass casting for joining the halves of glass molds, making items such as bottles and jars. Welding glass is accomplished by heating the glass through the glass transition, turning it into a thick, formable, liquid mass. Heating is usually done with a gas or oxy-gas torch, or a furnace, because the temperatures for melting glass are often quite high. This temperature may vary, depending on the type of glass. For example, lead glass becomes a weldable liquid at around 1,600 °F (870 °C), and can be welded with a simple propane torch. On the other hand, quartz glass (fused silica) must be heated to over 3,000 °F (1,650 °C), but quickly loses its viscosity and formability if overheated, so an oxyhydrogen torch must be used. Sometimes a tube may be attached to the glass, allowing it to be blown into various shapes, such as bulbs, bottles, or tubes. When two pieces of liquid glass are pressed together, they will usually weld very readily. Welding a handle onto a pitcher can usually be done with relative ease. However, when welding a tube to another tube, a combination of blowing and suction, and pressing and pulling is used to ensure a good seal, to shape the glass, and to keep the surface tension from closing the tube in on itself. Sometimes a filler rod may be used, but usually not.

Because glass is very brittle in its solid state, it is often prone to cracking upon heating and cooling, especially if the heating and cooling are uneven. This is because the brittleness of glass does not allow for uneven thermal expansion. Glass that has been welded will usually need to be cooled very slowly and evenly through the glass transition, in a process called annealing, to relieve any internal stresses created by a temperature gradient.

There are many types of glass, and it is most common to weld using the same types. Different glasses often have different rates of thermal expansion, which can cause them to crack upon cooling when they contract differently. For instance, quartz has very low thermal expansion, while soda-lime glass has very high thermal expansion. When welding different glasses to each other, it is usually important to closely match their coefficients of thermal expansion, to ensure that cracking does not occur. Also, some glasses will simply not mix with others, so welding between certain types may not be possible.

Glass can also be welded to metals and ceramics, although with metals the process is usually more adhesion to the surface of the metal rather than a commingling of the two materials. However, certain glasses will typically bond only to certain metals. For example, lead glass bonds readily to copper or molybdenum, but not to aluminum. Tungsten electrodes are often used in lighting but will not bond to quartz glass, so the tungsten is often wetted with molten borosilicate glass, which bonds to both tungsten and quartz. However, care must be taken to ensure that all materials have similar coefficients of thermal expansion to prevent cracking both when the object cools and when it is heated again. Special alloys are often used for this purpose, ensuring that the coefficients of expansion match, and sometimes thin, metallic coatings may be applied to a metal to create a good bond with the glass.

Plastic Welding

Plastics are generally divided into two categories, which are "thermosets" and "thermoplastics." A thermoset is a plastic in which a chemical reaction sets the molecular bonds after first forming the plastic, and then the bonds cannot be broken again without degrading the plastic. Thermosets cannot be melted, therefore, once a thermoset has set it is impossible to weld it. Examples of thermosets include epoxies, silicone, vulcanized rubber, polyester, and polyurethane.

Thermoplastics, by contrast, form long molecular chains, which are often coiled or intertwined, forming an amorphous structure without any long-range, crystalline order. Some thermoplastics may be fully amorphous, while others have a partially crystalline/partially amorphous structure. Both amorphous and semicrystalline thermoplastics have a glass transition, above which welding can occur, but semicrystallines also have a specific melting point which is above the glass transition. Above this melting point, the viscous liquid will become a free-flowing liquid. Examples of thermoplastics include polyethylene, polypropylene, polystyrene, polyvinylchloride (PVC), and fluoroplastics like Teflon and Spectralon.

Welding thermoplastic is very similar to welding glass. The plastic first must be cleaned and then heated through the glass transition, turning the weld-interface into a thick, viscous liquid. Two heated interfaces can then be pressed together, allowing the molecules to mix through intermolecular diffusion, joining them as one. Then the plastic is cooled through the glass transition, allowing the weld to solidify. A filler rod may often be used for certain types of joints. The main differences between welding glass and plastic are the types of heating methods, the much lower melting temperatures, and the fact that plastics will burn if overheated. Many different methods have been devised for heating plastic to a weldable temperature without burning it. Ovens or electric heating tools can be used to melt the plastic. Ultrasonic, laser, or friction heating are other methods. Resistive metals may be implanted in the plastic, which respond to induction heating. Some plastics will begin to burn at temperatures lower than their glass transition, so welding can be performed by blowing a heated, inert gas onto the plastic, melting it while, at the same time, shielding it from oxygen.

Many thermoplastics can also be welded using chemical solvents. When placed in contact with the plastic, the solvent will begin to soften it, bringing the surface into a thick, liquid solution. When two melted surfaces are pressed together, the molecules in the solution mix, joining them as one. Because the solvent can permeate the plastic, the solvent evaporates out through the surface of the plastic, causing the weld to drop out of solution and solidify. A common use for solvent welding is for joining PVC or ABS (acrylonitrile butadiene styrene) pipes during plumbing, or for welding styrene and polystyrene plastics in the construction of models. Solvent welding is especially effective on plastics like PVC which burn at or below their glass transition, but may be ineffective on plastics like Teflon or polyethylene that are resistant to chemical decomposition.

Induction Welding

Induction welding is a form of welding that uses electromagnetic induction to heat the workpiece. The welding apparatus contains an induction coil that is energised with a radio-frequency electric current. This generates a high-frequency electromagnetic field that acts on either an electrically conductive or a ferromagnetic workpiece. In an electrically conductive workpiece, the main heating effect is resistive heating, which is due to induced currents called eddy currents. In a ferromagnetic workpiece, the heating is caused mainly by hysteresis, as the electromagnetic field repeatedly distorts the magnetic domains of the ferromagnetic material. In practice, most materials undergo a combination of these two effects.

Nonmagnetic materials and electrical insulators such as plastics can be induction-welded by implanting them with metallic or ferromagnetic compounds, called susceptors, that absorb the electromagnetic energy from the induction coil, become hot, and lose their heat to the surrounding material by thermal conduction. Plastic can also be induction welded by embedding the plastic with electrically conductive fibers like metals or carbon fiber. Induced eddy currents resistively heat the embedded fibers which lose their heat to the surrounding plastic by conduction. Induction welding of carbon fiber reinforced plastics is commonly used in the aerospace industry.

Induction welding is used for long production runs and is a highly automated process, usually used for welding the seams of pipes. It can be a very fast process, as a lot of power can be transferred to a localised area, so the faying surfaces melt very quickly and can be pressed together to form a continuous rolling weld.

The depth that the currents, and therefore heating, penetrates from the surface is inversely proportional to the square root of the frequency. The temperature of the metals being welded and their composition will also affect the penetration depth. This process is very similar to resistance welding, except that in the case of resistance welding the current is delivered using contacts to the workpiece instead of using induction.

Welder

A welder or lit operator is a tradesperson who specializes in fusing materials together. The term welder refers to the operator, the machine is referred to as the welding power supply. The materials to be joined can be metals (such as steel, aluminum, brass, stainless steel etc.) or varieties of plastic or polymer. Welders typically have to have good dexterity and attention to detail, as well as some technical knowledge about the materials being joined and best practices in the field.

Safety Issues

Welding, without the proper precautions appropriate for the process, can be a dangerous and unhealthy practice. However, with the use of new technology and proper protection, the risks of injury and death associated with welding can be greatly reduced. Because many common welding procedures involve an open electric arc or flame, the risk of burns is significant. To prevent them, welders wear personal protective equipment in the form of heavy leather gloves and protective long sleeve jackets to avoid exposure to extreme heat and flames. Additionally, the brightness of the weld area leads to a condition called arc eye in which ultraviolet light causes the inflammation of the cornea and can burn the retinas of the eyes. Full face welding helmets with dark face plates are worn to prevent this exposure, and in recent years, new helmet models have been produced that feature a face plate that self-darkens upon exposure to high amounts of UV light. To protect bystanders, opaque welding curtains often surround the welding area. These curtains, made of a polyvinyl chloride plastic film, shield nearby workers from exposure to the UV light from the electric arc, but should not be used to replace the filter glass used in helmets.

Welders are also often exposed to dangerous gases and particulate matter. Processes like flux-cored arc welding and shielded metal arc welding produce smoke containing particles of various types of oxides, which in some cases can lead to medical conditions like metal fume fever. The size of the particles in question tends to influence the toxicity of the fumes, with smaller particles presenting greater danger. Additionally, many processes produce fumes and various gases, most commonly carbon dioxide and ozone, that can prove dangerous if ventilation is inadequate. Furthermore, because the use of compressed gases and flames in many welding processes pose an explosion and fire risk, some common precautions include limiting the amount of oxygen in the air and keeping combustible materials away from the workplace. Welders with expertise in welding pressurized vessels, including submarine hulls, industrial boilers, and power plant heat exchangers and boilers, are generally referred to as boilermakers.

Notable Welders

Notable people who have worked as welders include:

- İshak Alaton, Turkish businessman and investor

- Lucian Boz, Romanian literary critic, essayist, novelist, poet and translator

- Bevan Braithwaite, chief executive of The Welding Institute

- Hardcore Holly, American semi-retired professional wrestler

- Mark Honadel, American businessman, former professional metal fabricator, welding instructor, industrial manager and politician

- William A. Schmidt, American welder, shop foreman and politician

- Stefan Löfven, Prime Minister of Sweden

- Werner Herzog, German film director

- Honoré Sharrer, American painter

References

- Mel Schwartz (2011). Innovations in Materials Manufacturing, Fabrication, and Environmental Safety. CRC Press. pp. 300–. ISBN 978-1-4200-8215-9. Retrieved 10 July 2012.

- AWS A3.0:2001, Standard Welding Terms and Definitions Including Terms for Adhesive Bonding, Brazing, Soldering, Thermal Cutting, and Thermal Spraying, American Welding Society (2001), p. 117. ISBN 0-87171-624-0

- ASM International (2003). Trends in Welding Research. Materials Park, Ohio: ASM International. pp. 995–1005. ISBN 0-87170-780-2.

- Freek Bos, Christian Louter, Fred Veer (2008) Challenging Glass: Conference on Architectural and Structural Applications. JOS Press. p. 194. ISBN 1586038664

- Cary, Howard B; Helzer, Scott C. (2005). Modern Welding Technology. Upper Saddle River, New Jersey: Pearson Education. ISBN 0-13-113029-3.

- Cary, Howard B. and Scott C. Helzer (2005). Modern Welding Technology. Upper Saddle River, New Jersey: Pearson Education. ISBN 0-13-113029-3.

- Blunt, Jane and Nigel C. Balchin (2002). Health and Safety in Welding and Allied Processes. Cambridge: Woodhead. ISBN 1-85573-538-5.

- Babini, A; Forzan (January 2002). "Eddy Current Distribution in a Thin Aluminum Layer" (PDF). Flux Magazine (38): 11–12. Retrieved 9 Mar 2015.

Methods of Welding

Shielded metal arc welding is a manual method of welding which uses electrodes covered with a flux to lay the weld. The alternative methods of welding are gas tungsten arc welding, flux-cored arc welding, submerged arc welding, spot welding, forge welding and robot welding. This section discusses the methods of welding in a critical manner providing key analysis to the subject matter.

Shielded Metal Arc Welding

Shielded metal arc welding (SMAW), also known as manual metal arc welding (MMA or MMAW), flux shielded arc welding or informally as stick welding, is a manual arc welding process that uses a consumable electrode covered with a flux to lay the weld.

Shielded metal arc welding

An electric current, in the form of either alternating current or direct current from a welding power supply, is used to form an electric arc between the electrode and the metals to be joined. The workpiece and the electrode melts forming a pool of molten metal (weld pool) that cools to form a joint. As the weld is laid, the flux coating of the electrode disintegrates, giving off vapors that serve as a shielding gas and providing a layer of slag, both of which protect the weld area from atmospheric contamination.

Because of the versatility of the process and the simplicity of its equipment and operation, shielded metal arc welding is one of the world's first and most popular welding processes. It dominates other welding processes in the maintenance and repair industry, and though flux-cored arc welding is growing in popularity, SMAW continues to be used extensively in the construction of heavy steel structures and in industrial fabrication. The process is used primarily to weld iron and steels (including stainless steel) but aluminium, nickel and copper alloys can also be welded with this method.

Development

After the discovery of the short pulsed electric arc in 1800 by Humphry Davy and of the continuous electric arc in 1802 by Vasily Petrov, there was little development in electrical welding until Auguste de Méritens developed a carbon arc torch that was patented in 1881.

In 1885, Nikolay Benardos and Stanisław Olszewski developed carbon arc welding, obtaining American patents from 1887 showing a rudimentary electrode holder. In 1888, the consumable metal electrode was invented by Nikolay Slavyanov. Later in 1890, C. L. Coffin received U.S. Patent 428,459 for his arc welding method that utilized a metal electrode. The process, like SMAW, deposited melted electrode metal into the weld as filler.

Around 1900, A. P. Strohmenger and Oscar Kjellberg released the first coated electrodes. Strohmenger used clay and lime coating to stabilize the arc, while Kjellberg dipped iron wire into mixtures of carbonates and silicates to coat the electrode. In 1912, Strohmenger released a heavily coated electrode, but high cost and complex production methods prevented these early electrodes from gaining popularity. In 1927, the development of an extrusion process reduced the cost of coating electrodes while allowing manufacturers to produce more complex coating mixtures designed for specific applications. In the 1950s, manufacturers introduced iron powder into the flux coating, making it possible to increase the welding speed.

In 1938 K. K. Madsen described an automated variation of SMAW, now known as gravity welding. It briefly gained popularity in the 1960s after receiving publicity for its use in Japanese shipyards though today its applications are limited. Another little used variation of the process, known as firecracker welding, was developed around the same time by George Hafergut in Austria.

Operation

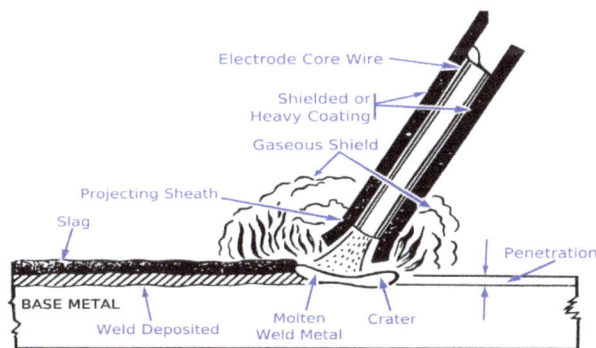

SMAW weld area

To strike the electric arc, the electrode is brought into contact with the workpiece by a very light touch with the electrode to the base metal then is pulled back slightly. This initiates the arc and thus the melting of the workpiece and the consumable electrode, and causes droplets of the electrode to be passed from the electrode to the weld pool. Striking an arc, which varies widely based upon electrode and workpiece composition, can be the hardest skill for beginners. The orientation of the electrode to workpiece is where most stumble, if the electrode is held at a perpendicular angle to the workpiece the tip will likely stick to the metal which will fuse the electrode to the workpiece which will cause it to heat up very rapidly. The tip of the electrode needs to be at a lower angle to the workpiece, which allows the weld pool to flow out of the arc. As the electrode melts, the flux covering disintegrates, giving off shielding gases that protect the weld area from oxygen and other atmospheric gases. In addition, the flux provides molten slag which covers the filler metal as it travels from the electrode to the weld pool. Once part of the weld pool, the slag floats to the surface and protects the weld from contamination as it solidifies. Once hardened, it must be chipped away to reveal the finished weld. As welding progresses and the electrode melts, the welder must periodically stop welding to remove the remaining electrode stub and insert a new electrode into the electrode holder. This activity, combined with chipping away the slag, reduces the amount of time that the welder can spend laying the weld, making SMAW one of the least efficient welding processes. In general, the operator factor, or the percentage of operator's time spent laying weld, is approximately 25%.

The actual welding technique utilized depends on the electrode, the composition of the workpiece, and the position of the joint being welded. The choice of electrode and welding position also determine the welding speed. Flat welds require the least operator skill, and can be done with electrodes that melt quickly but solidify slowly. This permits higher welding speeds.

Sloped, vertical or upside-down welding requires more operator skill, and often necessitates the use of an electrode that solidifies quickly to prevent the molten metal from flowing out of the weld pool. However, this generally means that the electrode melts less quickly, thus increasing the time required to lay the weld.

Quality

The most common quality problems associated with SMAW include weld spatter, porosity, poor fusion, shallow penetration, and cracking.

Weld spatter, while not affecting the integrity of the weld, damages its appearance and increases cleaning costs. It can be caused by excessively high current, a long arc, or arc blow, a condition associated with direct current characterized by the electric arc being deflected away from the weld pool by magnetic forces. Arc blow can also cause porosity in the weld, as can joint contamination, high welding speed, and a long welding arc, especially when low-hydrogen electrodes are used.

Porosity, often not visible without the use of advanced nondestructive testing methods, is a serious concern because it can potentially weaken the weld. Another defect affecting the strength of the weld is poor fusion, though it is often easily visible. It is caused by low current, contaminated joint surfaces, or the use of an improper electrode.

Shallow penetration, another detriment to weld strength, can be addressed by decreasing welding speed, increasing the current or using a smaller electrode. Any of these weld-strength-related defects can make the weld prone to cracking, but other factors are involved as well. High carbon, alloy or sulfur content in the base material can lead to cracking, especially if low-hydrogen electrodes and preheating are not employed. Furthermore, the workpieces should not be excessively restrained, as this introduces residual stresses into the weld and can cause cracking as the weld cools and contracts.

Safety

SMAW welding, like other welding methods, can be a dangerous and unhealthy practice if proper precautions are not taken. The process uses an open electric arc, which presents a risk of burns which are prevented by personal protective equipment in the form of heavy leather gloves and long sleeve jackets. Additionally, the brightness of the weld area can lead to a condition called arc eye, in which ultraviolet light causes inflammation of the cornea and can burn the retinas of the eyes. Welding helmets with dark face plates are worn to prevent this exposure, and in recent years, new helmet models have been produced that feature a face plate that self-darkens upon exposure to high amounts of UV light. To protect bystanders, especially in industrial environments, translucent welding curtains often surround the welding area. These curtains, made of a polyvinyl chloride plastic film, shield nearby workers from exposure to the UV light from the electric arc, but should not be used to replace the filter glass used in helmets.

In addition, the vaporizing metal and flux materials expose welders to dangerous gases and particulate matter. The smoke produced contains particles of various types of oxides. The size of the particles in question tends to influence the toxicity of the fumes, with smaller particles presenting a greater danger. Additionally, gases like carbon dioxide and ozone can form, which can prove dangerous if ventilation is inadequate. Some of the latest welding masks are fitted with an electric powered fan to help disperse harmful fumes.

Application and Materials

Shielded metal arc welding is one of the world's most popular welding processes, accounting for over half of all welding in some countries. Because of its versatility and simplicity, it is particularly dominant in the maintenance and repair industry, and is heavily used in the construction of steel structures and in industrial fabrication. In recent years its use has declined as flux-cored arc welding has expanded in the construction industry and gas metal arc welding has become more popular in industrial environments. However, because of the low equipment cost and wide applicability, the process will likely remain popular, especially among amateurs and small businesses where specialized welding processes are uneconomical and unnecessary.

SMAW is often used to weld carbon steel, low and high alloy steel, stainless steel, cast iron, and ductile iron. While less popular for nonferrous materials, it can be used on nickel and copper and their alloys and, in rare cases, on aluminium. The thickness of the material being welded is bounded on the low end primarily by the skill of the welder, but rarely does it drop below 1.5 mm (0.06 in). No upper bound exists: with proper joint preparation and use of multiple passes, materials of virtually unlimited thicknesses can be joined. Furthermore, depending on the electrode used and the skill of the welder, SMAW can be used in any position.

Equipment

Shielded metal arc welding equipment typically consists of a constant current welding power supply and an electrode, with an electrode holder, a 'ground' clamp, and welding cables (also known as welding leads) connecting the two.

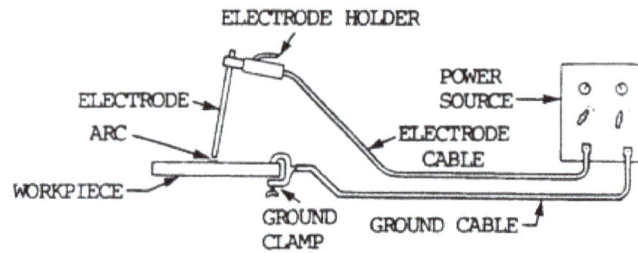

SMAW system setup

Power Supply

The power supply used in SMAW has constant current output, ensuring that the current (and thus the heat) remains relatively constant, even if the arc distance and voltage change. This is important because most applications of SMAW are manual, requiring that an operator hold the torch. Maintaining a suitably steady arc distance is difficult if a constant voltage power source is used instead, since it can cause dramatic heat variations and make welding more difficult. However, because the current is not maintained absolutely constant, skilled welders performing complicated welds can vary the arc length to cause minor fluctuations in the current.

A high output welding power supply for Stick, GTAW, MIG, Flux-Cored, & Gouging

The preferred polarity of the SMAW system depends primarily upon the electrode being used and the desired properties of the weld. Direct current with a negatively charged electrode (DCEN) causes heat to build up on the electrode, increasing the electrode melting rate and decreasing the depth of the weld. Reversing the polarity so that the electrode is positively charged (DCEP) and the workpiece is negatively charged increases the weld penetration. With alternating current the polarity changes over 100 times per second, creating an even heat distribution and providing a balance between electrode melting rate and penetration.

Typically, the equipment used for SMAW consists of a step-down transformer and for direct current models a rectifier, which converts alternating current into direct current. Because the power normally supplied to the welding machine is high-voltage alternating current, the welding transformer is used to reduce the voltage and increase the current. As a result, instead of 220 V at 50 A, for example, the power supplied by the transformer is around 17–45 V at currents up to 600 A. A number of different types of transformers can be used to produce this effect, including multiple coil and inverter machines, with each using a different method to manipulate the welding current. The multiple coil type adjusts the current by either varying the number of turns in the coil (in tap-type transformers) or by varying the distance between the primary and secondary coils (in movable coil or movable core transformers). Inverters, which are smaller and thus more portable, use electronic components to change the current characteristics.

Electrical generators and alternators are frequently used as portable welding power supplies, but because of lower efficiency and greater costs, they are less frequently used in industry. Maintenance also tends to be more difficult, because of the complexities of using a combustion engine as a power source. However, in one sense they are simpler: the use of a separate rectifier is unnecessary because they can provide either AC or DC. However, the engine driven units are most practical in field work where the welding often must be done out of doors and in locations where transformer type welders are not usable because there is no power source available to be transformed.

In some units the alternator is essentially the same as that used in portable generating sets used to supply mains power, modified to produce a higher current at a lower voltage but still at the 50 or 60 Hz grid frequency. In higher-quality units an alternator with more poles is used and supplies current at a higher frequency, such as 400 Hz. The smaller amount of time the high-frequency waveform spends near zero makes it much easier to strike and maintain a stable arc than with the cheaper grid-frequency sets or grid-frequency mains-powered units.

Electrode

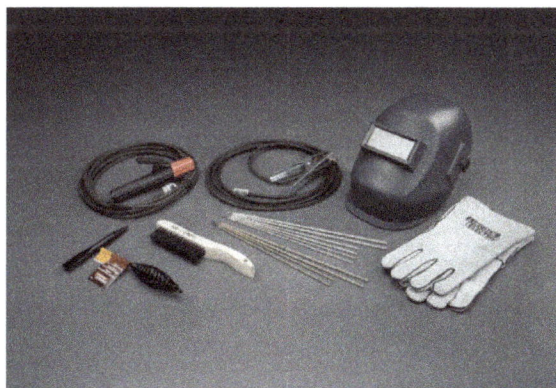

Various accessories for SMAW

The choice of electrode for SMAW depends on a number of factors, including the weld material, welding position and the desired weld properties. The electrode is coated in a metal mixture called flux, which gives off gases as it decomposes to prevent weld contamination, introduces deoxidizers to purify the weld, causes weld-protecting slag to form, improves the arc stability, and provides alloying elements to improve the weld quality. Electrodes can be divided into three groups—those designed to melt quickly are called "fast-fill" electrodes, those designed to solidify quickly are called "fast-freeze" electrodes, and intermediate electrodes go by the name "fill-freeze" or "fast-follow" electrodes. Fast-fill electrodes are designed to melt quickly so that the welding speed can be maximized, while fast-freeze electrodes supply filler metal that solidifies quickly, making welding in a variety of positions possible by preventing the weld pool from shifting significantly before solidifying.

The composition of the electrode core is generally similar and sometimes identical to that of the base material. But even though a number of feasible options exist, a slight difference in alloy composition can strongly impact the properties of the resulting weld. This is especially true of alloy steels such as HSLA steels. Likewise, electrodes of compositions similar to those of the base materials are often used for welding nonferrous materials like aluminium and copper. However, sometimes it is desirable to use electrodes with core materials significantly different from the base material. For example, stainless steel electrodes are sometimes used to weld two pieces of carbon steel, and are often utilized to weld stainless steel workpieces with carbon steel workpieces.

Electrode coatings can consist of a number of different compounds, including rutile, calcium fluoride, cellulose, and iron powder. Rutile electrodes, coated with 25%–45% TiO_2, are characterized by ease of use and good appearance of the resulting weld. However, they create welds with high hydrogen content, encouraging embrittlement and cracking. Electrodes containing calcium fluoride (CaF_2), sometimes known as basic or low-hydrogen electrodes, are hygroscopic and must be stored in dry conditions. They produce strong welds, but with a coarse and convex-shaped joint surface. Electrodes coated with cellulose, especially when combined with rutile, provide deep weld penetration, but because of their high moisture content, special procedures must be used to prevent excessive risk of cracking. Finally, iron powder is a common coating additive that increases the rate at which the electrode fills the weld joint, up to twice as fast.

To identify different electrodes, the American Welding Society established a system that assigns electrodes with a four- or five-digit number. Covered electrodes made of mild or low alloy steel carry the prefix E, followed by their number. The first two or three digits of the number specify the tensile strength of the weld metal, in thousand pounds per square inch (ksi). The penultimate digit generally identifies the welding positions permissible with the electrode, typically using the values 1 (normally fast-freeze electrodes, implying all position welding) and 2 (normally fast-fill electrodes, implying horizontal welding only). The welding current and type of electrode covering are specified by the last two digits together. When applicable, a suffix is used to denote the alloying element being contributed by the electrode.

Common electrodes include the E6010, a fast-freeze, all-position electrode with a minimum tensile strength of 60 ksi (410 MPa) which is operated using DCEP. E6011 is similar except its flux coating allows it to be used with alternating current in addition to DCEP. E7024 is a fast-fill electrode, used primarily to make flat or horizontal welds using AC, DCEN, or DCEP. Examples of fill-freeze electrodes are the E6012, E6013, and E7014, all of which provide a compromise between fast welding speeds and all-position welding.

Process Variations

Though SMAW is almost exclusively a manual arc welding process, one notable process variation exists, known as gravity welding or gravity arc welding. It serves as an automated version of the traditional shielded metal arc welding process, employing an electrode holder attached to an inclined bar along the length of the weld. Once started, the process continues until the electrode is spent, allowing the operator to manage multiple gravity welding systems. The electrodes employed (often E6027 or E7024) are coated heavily in flux, and are typically 71 cm (28 in) in length and about 6.35 mm (0.25 in) thick. As in manual SMAW, a constant current welding power supply is used, with either negative polarity direct current or alternating current. Due to a rise in the use of semiautomatic welding processes such as flux-cored arc welding, the popularity of gravity welding has fallen as its economic advantage over such methods is often minimal. Other SMAW-related methods that are even less frequently used include firecracker welding, an automatic method for making butt and fillet welds, and massive electrode welding, a process for welding large components or structures that can deposit up to 27 kg (60 lb) of weld metal per hour.

Gas Tungsten Arc Welding

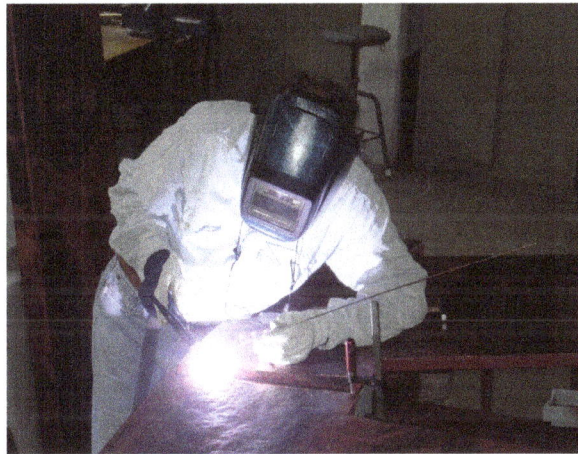

TIG welding of a bronze sculpture

Gas tungsten arc welding (GTAW), also known as tungsten inert gas (TIG) welding, is an arc welding process that uses a non-consumable tungsten electrode to produce the weld. The weld area is protected from atmospheric contamination by an inert shielding gas (argon or helium), and a filler metal is normally used, though some welds, known as autogenous welds, do not require it. A constant-current welding power supply produces electrical energy, which is conducted across the arc through a column of highly ionized gas and metal vapors known as a plasma.

GTAW is most commonly used to weld thin sections of stainless steel and non-ferrous metals such as aluminum, magnesium, and copper alloys. The process grants the operator greater control over the weld than competing processes such as shielded metal arc welding and gas metal arc welding, allowing for stronger, higher quality welds. However, GTAW is comparatively more complex and difficult to master, and furthermore, it is significantly slower than most other welding techniques.

A related process, plasma arc welding, uses a slightly different welding torch to create a more focused welding arc and as a result is often automated.

Development

After the discovery of the short pulsed electric arc in 1800 by Humphry Davy and of the continuous electric arc in 1802 by Vasily Petrov, arc welding developed slowly. C. L. Coffin had the idea of welding in an inert gas atmosphere in 1890, but even in the early 20th century, welding non-ferrous materials such as aluminum and magnesium remained difficult because these metals react rapidly with the air and result in porous, dross-filled welds. Processes using flux-covered electrodes did not satisfactorily protect the weld area from contamination. To solve the problem, bottled inert gases were used in the beginning of the 1930s. A few years later, a direct current, gas-shielded welding process emerged in the aircraft industry for welding magnesium.

Russell Meredith of Northrop Aircraft perfected the process in 1941. Meredith named the process Heliarc because it used a tungsten electrode arc and helium as a shielding gas, but it is often referred to as tungsten inert gas welding (TIG). The American Welding Society's official term is gas tungsten arc welding (GTAW). Linde Air Products developed a wide range of air-cooled and water-cooled torches, gas lenses to improve shielding, and other accessories that increased the use of the process. Initially, the electrode overheated quickly and, despite tungsten's high melting temperature, particles of tungsten were transferred to the weld. To address this problem, the polarity of the electrode was changed from positive to negative, but the change made it unsuitable for welding many non-ferrous materials. Finally, the development of alternating current units made it possible to stabilize the arc and produce high quality aluminum and magnesium welds.

Developments continued during the following decades. Linde developed water-cooled torches that helped prevent overheating when welding with high currents. During the 1950s, as the process continued to gain popularity, some users turned to carbon dioxide as an alternative to the more expensive welding atmospheres consisting of argon and helium, but this proved unacceptable for welding aluminum and magnesium because it reduced weld quality, so it is rarely used with GTAW today. The use of any shielding gas containing an oxygen compound, such as carbon dioxide, quickly contaminates the tungsten electrode, making it unsuitable for the TIG process. In 1953, a new process based on GTAW was developed, called plasma arc welding. It affords greater control and improves weld quality by using a nozzle to focus the electric arc, but is largely limited to automated systems, whereas GTAW remains primarily a manual, hand-held method. Development within the GTAW process has continued as well, and today a number of variations exist. Among the most popular are the pulsed-current, manual programmed, hot-wire, dabber, and increased penetration GTAW methods.

Operation

Manual gas tungsten arc welding is a relatively difficult welding method, due to the coordination required by the welder. Similar to torch welding, GTAW normally requires two hands, since most applications require that the welder manually feed a filler metal into the weld area with one hand while manipulating the welding torch in the other. Maintaining a short arc length, while preventing contact between the electrode and the workpiece, is also important.

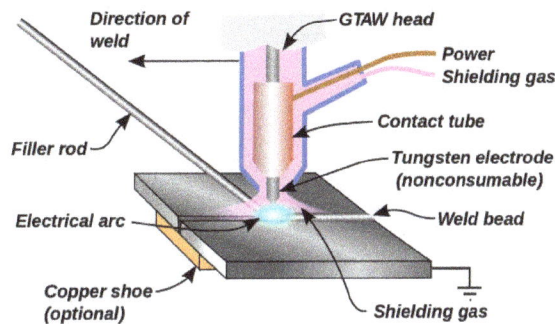

Direction of weld — GTAW head — Power — Shielding gas — Contact tube — Tungsten electrode (nonconsumable) — Weld bead — Filler rod — Electrical arc — Copper shoe (optional) — Shielding gas

GTAW weld area

To strike the welding arc, a high frequency generator (similar to a Tesla coil) provides an electric spark. This spark is a conductive path for the welding current through the shielding gas and allows the arc to be initiated while the electrode and the workpiece are separated, typically about 1.5–3 mm (0.06–0.12 in) apart.

Once the arc is struck, the welder moves the torch in a small circle to create a welding pool, the size of which depends on the size of the electrode and the amount of current. While maintaining a constant separation between the electrode and the workpiece, the operator then moves the torch back slightly and tilts it backward about 10–15 degrees from vertical. Filler metal is added manually to the front end of the weld pool as it is needed.

Welders often develop a technique of rapidly alternating between moving the torch forward (to advance the weld pool) and adding filler metal. The filler rod is withdrawn from the weld pool each time the electrode advances, but it is always kept inside the gas shield to prevent oxidation of its surface and contamination of the weld. Filler rods composed of metals with a low melting temperature, such as aluminum, require that the operator maintain some distance from the arc while staying inside the gas shield. If held too close to the arc, the filler rod can melt before it makes contact with the weld puddle. As the weld nears completion, the arc current is often gradually reduced to allow the weld crater to solidify and prevent the formation of crater cracks at the end of the weld.

Safety

Welders wear protective clothing, including light and thin leather gloves and protective long sleeve shirts with high collars, to avoid exposure to strong ultraviolet light. Due to the absence of smoke in GTAW, the electric arc light is not covered by fumes and particulate matter as in stick welding or shielded metal arc welding, and thus is a great deal brighter, subjecting operators to strong ultraviolet light. The welding arc has a different range and strength of UV light wavelengths from sunlight, but the welder is very close to the source and the light intensity is very strong. Potential arc light damage includes accidental flashes to the eye or arc eye and skin damage similar to strong sunburn. Operators wear opaque helmets with dark eye lenses and full head and neck coverage to prevent this exposure to UV light. Modern helmets often feature a liquid crystal-type face plate that self-darkens upon exposure to the bright light of the struck arc. Transparent welding curtains, made of a polyvinyl chloride plastic film, are often used to shield nearby workers and bystanders from exposure to the UV light from the electric arc.

Welders are also often exposed to dangerous gases and particulate matter. While the process doesn't produce smoke, the brightness of the arc in GTAW can break down surrounding air to form ozone and nitric oxides. The ozone and nitric oxides react with lung tissue and moisture to create nitric acid and ozone burn. Ozone and nitric oxide levels are moderate, but exposure duration, repeated exposure, and the quality and quantity of fume extraction, and air change in the room must be monitored. Welders who do not work safely can contract emphysema and oedema of the lungs, which can lead to early death. Similarly, the heat from the arc can cause poisonous fumes to form from cleaning and degreasing materials. Cleaning operations using these agents should not be performed near the site of welding, and proper ventilation is necessary to protect the welder.

Applications

While the aerospace industry is one of the primary users of gas tungsten arc welding, the process is used in a number of other areas. Many industries use GTAW for welding thin workpieces, especially nonferrous metals. It is used extensively in the manufacture of space vehicles, and is also frequently employed to weld small-diameter, thin-wall tubing such as those used in the bicycle industry. In addition, GTAW is often used to make root or first-pass welds for piping of various sizes. In maintenance and repair work, the process is commonly used to repair tools and dies, especially components made of aluminum and magnesium. Because the weld metal is not transferred directly across the electric arc like most open arc welding processes, a vast assortment of welding filler metal is available to the welding engineer. In fact, no other welding process permits the welding of so many alloys in so many product configurations. Filler metal alloys, such as elemental aluminum and chromium, can be lost through the electric arc from volatilization. This loss does not occur with the GTAW process. Because the resulting welds have the same chemical integrity as the original base metal or match the base metals more closely, GTAW welds are highly resistant to corrosion and cracking over long time periods, making GTAW the welding procedure of choice for critical operations like sealing spent nuclear fuel canisters before burial.

Quality

GTAW fillet weld

Gas tungsten arc welding, because it affords greater control over the weld area than other welding processes, can produce high-quality welds when performed by skilled operators. Maximum weld quality is assured by maintaining cleanliness—all equipment and materials used must be free from oil, moisture, dirt and other impurities, as these cause weld porosity and consequently a decrease in weld strength and quality. To remove oil and grease, alcohol or similar commercial solvents may be used, while a stainless steel wire brush or chemical process can remove oxides from the surfaces of metals like aluminum. Rust on steels can be removed by first grit blasting the surface and then using a wire brush to remove any embedded grit. These steps are especially important when negative polarity direct current is used, because such a power supply provides no cleaning during the welding process, unlike positive polarity direct current or alternating current. To maintain a clean weld pool during welding, the shielding gas flow should be sufficient and consistent so that the gas covers the weld and blocks impurities in the atmosphere. GTAW in windy or drafty environments increases the amount of shielding gas necessary to protect the weld, increasing the cost and making the process unpopular outdoors.

The level of heat input also affects weld quality. Low heat input, caused by low welding current or high welding speed, can limit penetration and cause the weld bead to lift away from the surface being welded. If there is too much heat input, however, the weld bead grows in width while the likelihood of excessive penetration and spatter increase. Additionally, if the welding torch is too far from the workpiece the shielding gas becomes ineffective, causing porosity within the weld. This results in a weld with pinholes, which is weaker than a typical weld.

If the amount of current used exceeds the capability of the electrode, tungsten inclusions in the weld may result. Known as tungsten spitting, this can be identified with radiography and can be prevented by changing the type of electrode or increasing the electrode diameter. In addition, if the electrode is not well protected by the gas shield or the operator accidentally allows it to contact the molten metal, it can become dirty or contaminated. This often causes the welding arc to become unstable, requiring that the electrode be ground with a diamond abrasive to remove the impurity.

Equipment

GTAW torch with various electrodes, cups, collets and gas diffusers

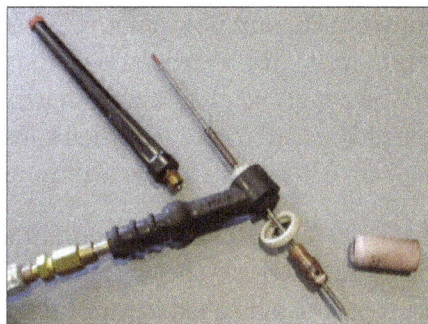

GTAW torch, disassembled

The equipment required for the gas tungsten arc welding operation includes a welding torch utilizing a non-consumable tungsten electrode, a constant-current welding power supply, and a shielding gas source.

Welding Torch

GTAW welding torches are designed for either automatic or manual operation and are equipped with cooling systems using air or water. The automatic and manual torches are similar in construction, but the manual torch has a handle while the automatic torch normally comes with a mounting rack. The angle between the centerline of the handle and the centerline of the tungsten electrode, known as the head angle, can be varied on some manual torches according to the preference of the operator. Air cooling systems are most often used for low-current operations (up to about 200 A), while water cooling is required for high-current welding (up to about 600 A). The torches are connected with cables to the power supply and with hoses to the shielding gas source and where used, the water supply.

The internal metal parts of a torch are made of hard alloys of copper or brass so it can transmit current and heat effectively. The tungsten electrode must be held firmly in the center of the torch with an appropriately sized collet, and ports around the electrode provide a constant flow of shielding gas. Collets are sized according to the diameter of the tungsten electrode they hold. The body of the torch is made of heat-resistant, insulating plastics covering the metal components, providing insulation from heat and electricity to protect the welder.

The size of the welding torch nozzle depends on the amount of shielded area desired. The size of the gas nozzle depends upon the diameter of the electrode, the joint configuration, and the availability of access to the joint by the welder. The inside diameter of the nozzle is preferably at least three times the diameter of the electrode, but there are no hard rules. The welder judges the effectiveness of the shielding and increases the nozzle size to increase the area protected by the external gas shield as needed. The nozzle must be heat resistant and thus is normally made of alumina or a ceramic material, but fused quartz, a high purity glass, offers greater visibility. Devices can be inserted into the nozzle for special applications, such as gas lenses or valves to improve the control shielding gas flow to reduce turbulence and introduction of contaminated atmosphere into the shielded area. Hand switches to control welding current can be added to the manual GTAW torches.

Power Supply

Gas tungsten arc welding uses a constant current power source, meaning that the current (and thus the heat) remains relatively constant, even if the arc distance and voltage change. This is important because most applications of GTAW are manual or semiautomatic, requiring that an operator hold the torch. Maintaining a suitably steady arc distance is difficult if a constant voltage power source is used instead, since it can cause dramatic heat variations and make welding more difficult.

The preferred polarity of the GTAW system depends largely on the type of metal being welded. Direct current with a negatively charged electrode (DCEN) is often employed when welding steels, nickel, titanium, and other metals. It can also be used in automatic GTAW of aluminum or magnesium when helium is used as a shielding gas. The negatively charged electrode generates heat by emitting electrons, which travel across the arc, causing thermal ionization of the shielding gas and increasing the temperature of the base material. The ionized shielding gas flows toward the electrode, not the base material, and this can allow oxides to build on the surface of the weld. Direct current with a positively charged electrode (DCEP) is less common, and is used primarily for shallow welds since less heat is generated in the base material. Instead of flowing from the electrode to the base material, as in DCEN, electrons go the other direction, causing the electrode to reach very high temperatures. To help it maintain its shape and prevent softening, a larger electrode is often used. As the electrons flow toward the electrode, ionized shielding gas flows back toward the base material, cleaning the weld by removing oxides and other impurities and thereby improving its quality and appearance.

GTAW power supply

Alternating current, commonly used when welding aluminum and magnesium manually or semi-automatically, combines the two direct currents by making the electrode and base material alternate between positive and negative charge. This causes the electron flow to switch directions constantly, preventing the tungsten electrode from overheating while maintaining the heat in the base material. Surface oxides are still removed during the electrode-positive portion of the cycle and the base metal is heated more deeply during the electrode-negative portion of the cycle. Some power supplies enable operators to use an unbalanced alternating current wave by modifying the exact percentage of time that the current spends in each state of polarity, giving them more control over the amount of heat and cleaning action supplied by the power source. In addition, operators must be wary of rectification, in which the arc fails to reignite as it passes from straight polarity (negative electrode) to reverse polarity (positive electrode). To remedy the problem, a square wave power supply can be used, as can high-frequency voltage to encourage ignition.

Electrode

ISO Class	ISO Color	AWS Class	AWS Color	Alloy
WP	Green	EWP	Green	None
WC20	Gray	EWCe-2	Orange	~2% CeO_2
WL10	Black	EWLa-1	Black	~1% La_2O_3
WL15	Gold	EWLa-1.5	Gold	~1.5% La_2O_3
WL20	Sky-blue	EWLa-2	Blue	~2% La_2O_3

WT10	Yellow	EWTh-1	Yellow	~1% ThO_2
WT20	Red	EWTh-2	Red	~2% ThO_2
WT30	Violet			~3% ThO_2
WT40	Orange			~4% ThO_2
WY20	Blue			~2% Y_2O_3
WZ3	Brown	EWZr-1	Brown	~0.3% ZrO_2
WZ8	White			~0.8% ZrO_2

The electrode used in GTAW is made of tungsten or a tungsten alloy, because tungsten has the highest melting temperature among pure metals, at 3,422 °C (6,192 °F). As a result, the electrode is not consumed during welding, though some erosion (called burn-off) can occur. Electrodes can have either a clean finish or a ground finish—clean finish electrodes have been chemically cleaned, while ground finish electrodes have been ground to a uniform size and have a polished surface, making them optimal for heat conduction. The diameter of the electrode can vary between 0.5 and 6.4 millimetres (0.02 and 0.25 in), and their length can range from 75 to 610 millimetres (3.0 to 24.0 in).

A number of tungsten alloys have been standardized by the International Organization for Standardization and the American Welding Society in ISO 6848 and AWS A5.12, respectively, for use in GTAW electrodes, and are summarized in the adjacent table.

- Pure tungsten electrodes (classified as WP or EWP) are general purpose and low cost electrodes. They have poor heat resistance and electron emission. They find limited use in AC welding of e.g. magnesium and aluminum.

- Cerium oxide (or ceria) as an alloying element improves arc stability and ease of starting while decreasing burn-off. Cerium addition is not as effective as thorium but works well, and cerium is not radioactive.

- An alloy of lanthanum oxide (or lanthana) has a similar effect as cerium, and is also not radioactive.

- Thorium oxide (or thoria) alloy electrodes offer excellent arc performance and starting, making them popular general purpose electrodes. However, it is somewhat radioactive, making inhalation of thorium vapors and dust a health risk, and disposal an environmental risk.

- Electrodes containing zirconium oxide (or zirconia) increase the current capacity while improving arc stability and starting and increasing electrode life.

Filler metals are also used in nearly all applications of GTAW, the major exception being the welding of thin materials. Filler metals are available with different diameters and are made of a variety of materials. In most cases, the filler metal in the form of a rod is added to the weld pool manually, but some applications call for an automatically fed filler metal, which often is stored on spools or coils.

Shielding Gas

As with other welding processes such as gas metal arc welding, shielding gases are necessary in

GTAW to protect the welding area from atmospheric gases such as nitrogen and oxygen, which can cause fusion defects, porosity, and weld metal embrittlement if they come in contact with the electrode, the arc, or the welding metal. The gas also transfers heat from the tungsten electrode to the metal, and it helps start and maintain a stable arc.

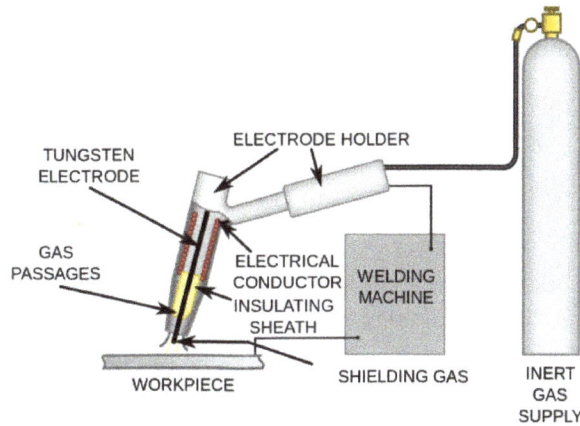

GTAW system setup

The selection of a shielding gas depends on several factors, including the type of material being welded, joint design, and desired final weld appearance. Argon is the most commonly used shielding gas for GTAW, since it helps prevent defects due to a varying arc length. When used with alternating current, argon shielding results in high weld quality and good appearance. Another common shielding gas, helium, is most often used to increase the weld penetration in a joint, to increase the welding speed, and to weld metals with high heat conductivity, such as copper and aluminum. A significant disadvantage is the difficulty of striking an arc with helium gas, and the decreased weld quality associated with a varying arc length.

Argon-helium mixtures are also frequently utilized in GTAW, since they can increase control of the heat input while maintaining the benefits of using argon. Normally, the mixtures are made with primarily helium (often about 75% or higher) and a balance of argon. These mixtures increase the speed and quality of the AC welding of aluminum, and also make it easier to strike an arc. Another shielding gas mixture, argon-hydrogen, is used in the mechanized welding of light gauge stainless steel, but because hydrogen can cause porosity, its uses are limited. Similarly, nitrogen can sometimes be added to argon to help stabilize the austenite in austenitic stainless steels and increase penetration when welding copper. Due to porosity problems in ferritic steels and limited benefits, however, it is not a popular shielding gas additive.

Materials

Gas tungsten arc welding is most commonly used to weld stainless steel and nonferrous materials, such as aluminum and magnesium, but it can be applied to nearly all metals, with a notable exception being zinc and its alloys. Its applications involving carbon steels are limited not because of process restrictions, but because of the existence of more economical steel welding techniques, such as gas metal arc welding and shielded metal arc welding. Furthermore, GTAW can be performed in a variety of other-than-flat positions, depending on the skill of the welder and the materials being welded.

Aluminum and Magnesium

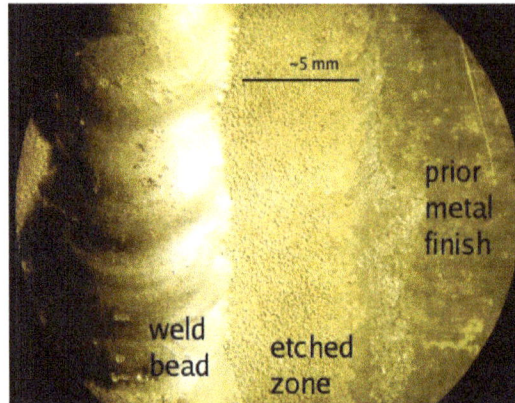

A TIG weld showing an accentuated AC etched zone

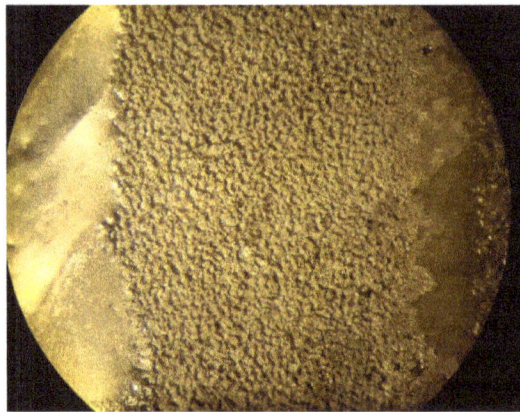

Closeup view of an aluminum TIG weld AC etch zone

Aluminum and magnesium are most often welded using alternating current, but the use of direct current is also possible, depending on the properties desired. Before welding, the work area should be cleaned and may be preheated to 175 to 200 °C (347 to 392 °F) for aluminum or to a maximum of 150 °C (302 °F) for thick magnesium workpieces to improve penetration and increase travel speed. AC current can provide a self-cleaning effect, removing the thin, refractory aluminum oxide (sapphire) layer that forms on aluminum metal within minutes of exposure to air. This oxide layer must be removed for welding to occur. When alternating current is used, pure tungsten electrodes or zirconiated tungsten electrodes are preferred over thoriated electrodes, as the latter are more likely to "spit" electrode particles across the welding arc into the weld. Blunt electrode tips are preferred, and pure argon shielding gas should be employed for thin workpieces. Introducing helium allows for greater penetration in thicker workpieces, but can make arc starting difficult.

Direct current of either polarity, positive or negative, can be used to weld aluminum and magnesium as well. Direct current with a negatively charged electrode (DCEN) allows for high penetration. Argon is commonly used as a shielding gas for DCEN welding of aluminum. Shielding gases with high helium contents are often used for higher penetration in thicker materials. Thoriated electrodes are suitable for use in DCEN welding of aluminum. Direct current with a positively charged electrode (DCEP) is used primarily for shallow welds, especially those with a joint thickness of less than 1.6 mm (0.063 in). A thoriated tungsten electrode is commonly used, along with a pure argon shielding gas.

Steels

For GTAW of carbon and stainless steels, the selection of a filler material is important to prevent excessive porosity. Oxides on the filler material and workpieces must be removed before welding to prevent contamination, and immediately prior to welding, alcohol or acetone should be used to clean the surface. Preheating is generally not necessary for mild steels less than one inch thick, but low alloy steels may require preheating to slow the cooling process and prevent the formation of martensite in the heat-affected zone. Tool steels should also be preheated to prevent cracking in the heat-affected zone. Austenitic stainless steels do not require preheating, but martensitic and ferritic chromium stainless steels do. A DCEN power source is normally used, and thoriated electrodes, tapered to a sharp point, are recommended. Pure argon is used for thin workpieces, but helium can be introduced as thickness increases.

Dissimilar Metals

Welding dissimilar metals often introduces new difficulties to GTAW welding, because most materials do not easily fuse to form a strong bond. However, welds of dissimilar materials have numerous applications in manufacturing, repair work, and the prevention of corrosion and oxidation. In some joints, a compatible filler metal is chosen to help form the bond, and this filler metal can be the same as one of the base materials (for example, using a stainless steel filler metal with stainless steel and carbon steel as base materials), or a different metal (such as the use of a nickel filler metal for joining steel and cast iron). Very different materials may be coated or "buttered" with a material compatible with a particular filler metal, and then welded. In addition, GTAW can be used in cladding or overlaying dissimilar materials.

When welding dissimilar metals, the joint must have an accurate fit, with proper gap dimensions and bevel angles. Care should be taken to avoid melting excessive base material. Pulsed current is particularly useful for these applications, as it helps limit the heat input. The filler metal should be added quickly, and a large weld pool should be avoided to prevent dilution of the base materials.

Process Variations

Pulsed-current

In the pulsed-current mode, the welding current rapidly alternates between two levels. The higher current state is known as the pulse current, while the lower current level is called the background current. During the period of pulse current, the weld area is heated and fusion occurs. Upon dropping to the background current, the weld area is allowed to cool and solidify. Pulsed-current GTAW has a number of advantages, including lower heat input and consequently a reduction in distortion and warpage in thin workpieces. In addition, it allows for greater control of the weld pool, and can increase weld penetration, welding speed, and quality. A similar method, manual programmed GTAW, allows the operator to program a specific rate and magnitude of current variations, making it useful for specialized applications.

Dabber

The dabber variation is used to precisely place weld metal on thin edges. The automatic process replicates the motions of manual welding by feeding a cold filler wire into the weld area and dabbing (or oscillating) it into the welding arc. It can be used in conjunction with pulsed current, and is used to weld a variety of alloys, including titanium, nickel, and tool steels. Common applications include rebuilding seals in jet engines and building up saw blades, milling cutters, drill bits, and mower blades.

Gas Metal Arc Welding

Gas metal arc welding (GMAW), sometimes referred to by its subtypes metal inert gas (MIG) welding or metal active gas (MAG) welding, is a welding process in which an electric arc forms between a consumable wire electrode and the workpiece metal(s), which heats the workpiece metal(s), causing them to melt and join.

Along with the wire electrode, a shielding gas feeds through the welding gun, which shields the process from contaminants in the air. The process can be semi-automatic or automatic. A constant voltage, direct current power source is most commonly used with GMAW, but constant current systems, as well as alternating current, can be used. There are four primary methods of metal transfer in GMAW, called globular, short-circuiting, spray, and pulsed-spray, each of which has distinct properties and corresponding advantages and limitations.

Originally developed for welding aluminium and other non-ferrous materials in the 1940s, GMAW was soon applied to steels because it provided faster welding time compared to other welding processes. The cost of inert gas limited its use in steels until several years later, when the use of semi-inert gases such as carbon dioxide became common. Further developments during the 1950s and 1960s gave the process more versatility and as a result, it became a highly used industrial process. Today, GMAW is the most common industrial welding process, preferred for its versatility, speed and the relative ease of adapting the process to robotic automation. Unlike welding processes that do not employ a shielding gas, such as shielded metal arc welding, it is rarely used outdoors or in other areas of air volatility. A related process, flux cored arc welding, often does not use a shielding gas, but instead employs an electrode wire that is hollow and filled with flux.

Development

The principles of gas metal arc welding began to be understood in the early 19th century, after Humphry Davy discovered the short pulsed electric arcs in 1800. Vasily Petrov independently produced the continuous electric arc in 1802 (followed by Davy after 1808). It was not until the 1880s that the technology became developed with the aim of industrial usage. At first, carbon electrodes were used in carbon arc welding. By 1890, metal electrodes had been invented by Nikolay Slavyanov and C. L. Coffin. In 1920, an early predecessor of GMAW was invented by P. O. Nobel of General Electric. It used a bare electrode wire and direct current, and used arc voltage to regulate the feed rate. It did not use a shielding gas to protect the weld, as developments in welding atmospheres did not take place until later that decade. In 1926 another forerunner of GMAW was released, but it was not suitable for practical use.

In 1948, GMAW was developed by the Battelle Memorial Institute. It used a smaller diameter electrode and a constant voltage power source developed by H. E. Kennedy. It offered a high deposition rate, but the high cost of inert gases limited its use to non-ferrous materials and prevented cost savings. In 1953, the use of carbon dioxide as a welding atmosphere was developed, and it quickly gained popularity in GMAW, since it made welding steel more economical. In 1958 and 1959, the short-arc variation of GMAW was released, which increased welding versatility and made the welding of thin materials possible while relying on smaller electrode wires and more advanced power supplies. It quickly became the most popular GMAW variation.

The spray-arc transfer variation was developed in the early 1960s, when experimenters added small amounts of oxygen to inert gases. More recently, pulsed current has been applied, giving rise to a new method called the pulsed spray-arc variation.

GMAW is one of the most popular welding methods, especially in industrial environments. It is used extensively by the sheet metal industry and, by extension, the automobile industry. There, the method is often used for arc spot welding, thereby replacing riveting or resistance spot welding. It is also popular for automated welding, in which robots handle the workpieces and the welding gun to speed up the manufacturing process. GMAW can be difficult to perform well outdoors, since drafts can dissipate the shielding gas and allow contaminants into the weld; flux cored arc welding is better suited for outdoor use such as in construction. Likewise, GMAW's use of a shielding gas does not lend itself to underwater welding, which is more commonly performed via shielded metal arc welding, flux cored arc welding, or gas tungsten arc welding.

Equipment

To perform gas metal arc welding, the basic necessary equipment is a welding gun, a wire feed unit, a welding power supply, a welding electrode wire, and a shielding gas supply.

Welding Gun and Wire Feed Unit

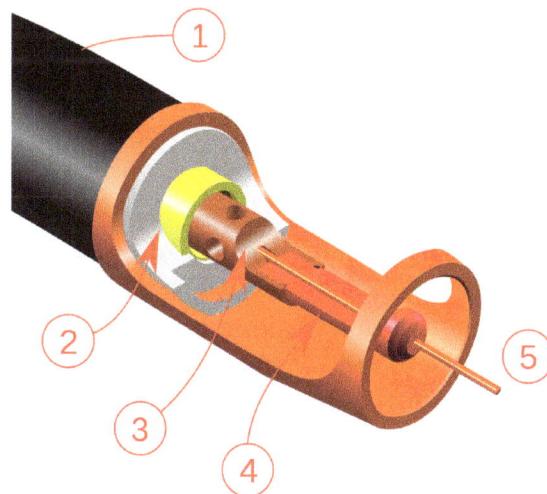

GMAW torch nozzle cutaway image. (1) Torch handle, (2) Molded phenolic dielectric (shown in white) and threaded metal nut insert (yellow), (3) Shielding gas diffuser, (4) Contact tip, (5) Nozzle output face

GMAW on stainless steel

The typical GMAW welding gun has a number of key parts—a control switch, a contact tip, a power cable, a gas nozzle, an electrode conduit and liner, and a gas hose. The control switch, or trigger, when pressed by the operator, initiates the wire feed, electric power, and the shielding gas flow, causing an electric arc to be struck. The contact tip, normally made of copper and sometimes chemically treated to reduce spatter, is connected to the welding power source through the power cable and transmits the electrical energy to the electrode while directing it to the weld area. It must be firmly secured and properly sized, since it must allow the electrode to pass while maintaining electrical contact. On the way to the contact tip, the wire is protected and guided by the electrode conduit and liner, which help prevent buckling and maintain an uninterrupted wire feed. The gas nozzle directs the shielding gas evenly into the welding zone. Inconsistent flow may not adequately protect the weld area. Larger nozzles provide greater shielding gas flow, which is useful for high current welding operations that develop a larger molten weld pool. A gas hose from the tanks of shielding gas supplies the gas to the nozzle. Sometimes, a water hose is also built into the welding gun, cooling the gun in high heat operations.

The wire feed unit supplies the electrode to the work, driving it through the conduit and on to the contact tip. Most models provide the wire at a constant feed rate, but more advanced machines can vary the feed rate in response to the arc length and voltage. Some wire feeders can reach feed rates as high as 30.5 m/min (1200 in/min), but feed rates for semiautomatic GMAW typically range from 2 to 10 m/min (75 – 400 in/min).

Tool Style

The most common electrode holder is a semiautomatic air-cooled holder. Compressed air circulates through it to maintain moderate temperatures. It is used with lower current levels for welding lap or butt joints. The second most common type of electrode holder is semiautomatic water-cooled, where the only difference is that water takes the place of air. It uses higher current levels for welding T or corner joints. The third typical holder type is a water cooled automatic electrode holder—which is typically used with automated equipment.

Power Supply

Most applications of gas metal arc welding use a constant voltage power supply. As a result, any change in arc length (which is directly related to voltage) results in a large change in heat input and current. A shorter arc length causes a much greater heat input, which makes the wire electrode melt more quickly and thereby restore the original arc length. This helps operators keep the arc length consistent even when manually welding with hand-held welding guns. To achieve a similar effect, sometimes a constant current power source is used in combination with an arc voltage-controlled wire feed unit. In this case, a change in arc length makes the wire feed rate adjust to maintain a relatively constant arc length. In rare circumstances, a constant current power source and a constant wire feed rate unit might be coupled, especially for the welding of metals with high thermal conductivities, such as aluminum. This grants the operator additional control over the heat input into the weld, but requires significant skill to perform successfully.

Alternating current is rarely used with GMAW; instead, direct current is employed and the electrode is generally positively charged. Since the anode tends to have a greater heat concentration, this results in faster melting of the feed wire, which increases weld penetration and welding speed. The polarity can be reversed only when special emissive-coated electrode wires are used, but since these are not popular, a negatively charged electrode is rarely employed.

Electrode

Electrode selection is based primarily on the composition of the metal being welded, the process variation being used, joint design and the material surface conditions. Electrode selection greatly influences the mechanical properties of the weld and is a key factor of weld quality. In general the finished weld metal should have mechanical properties similar to those of the base material with no defects such as discontinuities, entrained contaminants or porosity within the weld. To achieve these goals a wide variety of electrodes exist. All commercially available electrodes contain deoxidizing metals such as silicon, manganese, titanium and aluminum in small percentages to help prevent oxygen porosity. Some contain denitriding metals such as titanium and zirconium to avoid nitrogen porosity. Depending on the process variation and base material being welded the diameters of the electrodes used in GMAW typically range from 0.7 to 2.4 mm (0.028 – 0.095 in) but can be as large as 4 mm (0.16 in). The smallest electrodes, generally up to 1.14 mm (0.045 in) are associated with the short-circuiting metal transfer process, while the most common spray-transfer process mode electrodes are usually at least 0.9 mm (0.035 in).

Shielding Gas

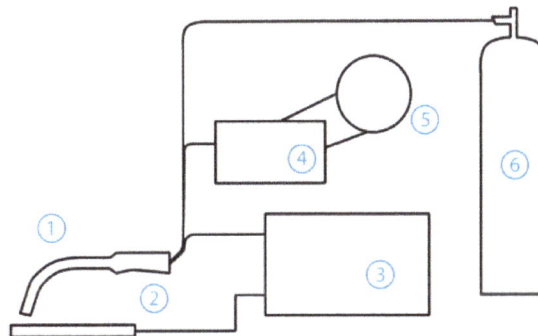

GMAW Circuit diagram. (1) Welding torch, (2) Workpiece, (3) Power source, (4) Wire feed unit, (5) Electrode source, (6) Shielding gas supply.

Shielding gases are necessary for gas metal arc welding to protect the welding area from atmospheric gases such as nitrogen and oxygen, which can cause fusion defects, porosity, and weld metal embrittlement if they come in contact with the electrode, the arc, or the welding metal. This problem is common to all arc welding processes; for example, in the older Shielded-Metal Arc Welding process (SMAW), the electrode is coated with a solid flux which evolves a protective cloud of carbon dioxide when melted by the arc. In GMAW, however, the electrode wire does not have a flux coating, and a separate shielding gas is employed to protect the weld. This eliminates slag, the hard residue from the flux that builds up after welding and must be chipped off to reveal the completed weld.

The choice of a shielding gas depends on several factors, most importantly the type of material being welded and the process variation being used. Pure inert gases such as argon and helium are only used for nonferrous welding; with steel they do not provide adequate weld penetration (argon) or cause an erratic arc and encourage spatter (with helium). Pure carbon dioxide, on the other hand, allows for deep penetration welds but encourages oxide formation, which adversely affect the mechanical properties of the weld. Its low cost makes it an attractive choice, but because of the reactivity of the arc plasma, spatter is unavoidable and welding thin materials is difficult. As a result, argon and carbon dioxide are frequently mixed in a 75%/25% to 90%/10% mixture. Generally, in short circuit GMAW, higher carbon dioxide content increases the weld heat and energy when all other weld parameters (volts, current, electrode type and diameter) are held the same. As the carbon dioxide content increases over 20%, spray transfer GMAW becomes increasingly problematic, especially with smaller electrode diameters.

Argon is also commonly mixed with other gases, oxygen, helium, hydrogen and nitrogen. The addition of up to 5% oxygen (like the higher concentrations of carbon dioxide mentioned above) can be helpful in welding stainless steel, however, in most applications carbon dioxide is preferred. Increased oxygen makes the shielding gas oxidize the electrode, which can lead to porosity in the deposit if the electrode does not contain sufficient deoxidizers. Excessive oxygen, especially when used in application for which it is not prescribed, can lead to brittleness in the heat affected zone. Argon-helium mixtures are extremely inert, and can be used on nonferrous materials. A helium concentration of 50–75% raises the required voltage and increases the heat in the arc, due to helium's higher ionization temperature. Hydrogen is sometimes added to argon in small concentrations (up to about 5%) for welding nickel and thick stainless steel workpieces. In higher concentrations (up to 25% hydrogen), it may be used for welding conductive materials such as copper. However, it should not be used on steel, aluminum or magnesium because it can cause porosity and hydrogen embrittlement.

Shielding gas mixtures of three or more gases are also available. Mixtures of argon, carbon dioxide and oxygen are marketed for welding steels. Other mixtures add a small amount of helium to argon-oxygen combinations, these mixtures are claimed to allow higher arc voltages and welding speed. Helium also sometimes serves as the base gas, with small amounts of argon and carbon dioxide added. However, because it is less dense than air, helium is less effective at shielding the weld than argon—which is denser than air. It also can lead to arc stability and penetration issues, and increased spatter, due to its much more energetic arc plasma. Helium is also substantially more expensive than other shielding gases. Other specialized and often proprietary gas mixtures claim even greater benefits for specific applications.

The desirable rate of shielding-gas flow depends primarily on weld geometry, speed, current, the type of gas, and the metal transfer mode. Welding flat surfaces requires higher flow than welding grooved materials, since gas disperses more quickly. Faster welding speeds, in general, mean that more gas must be supplied to provide adequate coverage. Additionally, higher current requires greater flow, and generally, more helium is required to provide adequate coverage than if argon is used. Perhaps most importantly, the four primary variations of GMAW have differing shielding gas flow requirements—for the small weld pools of the short circuiting and pulsed spray modes, about 10 L/min (20 ft³/h) is generally suitable, whereas for globular transfer, around 15 L/min (30 ft³/h) is preferred. The spray transfer variation normally requires more shielding-gas flow because of its higher heat input and thus larger weld pool. Typical gas-flow amounts are approximately 20–25 L/min (40–50 ft³/h).

Flux-cored Wire-fed

Flux-cored, self-shielding or gasless wire-fed welding had been developed for simplicity and portability. This avoids the gas system of conventional GMAW and uses a cored wire containing a solid flux. This flux vaporises during welding and produces a plume of shielding gas. Although described as a 'flux', this compound has little activity and acts mostly as an inert shield. The wire is of slightly larger diameter than for a comparable gas-shielded weld, to allow room for the flux. The smallest available is 0.8 mm diameter, compared to 0.6 mm for solid wire. The shield vapor is slightly active, rather than inert, so the process is always MAGS but not MIG (inert gas shield). This limits the process to steel and not aluminium.

Vaporising the additional flux requires greater heat in the wire, so these gasless machines operate as DCEP, rather than the DCEN usually used for GMAW to give deeper penetration. DCEP, or DC Electrode Positive, makes the welding wire into the positively-charged anode, which is the hotter side of the arc. Provided that it is switchable from DCEN to DCEP, a gas-shielded wire-feed machine may also be used for flux-cored wire.

Flux-cored wire is considered to have some advantages for outdoor welding on-site, as the shielding gas plume is less likely to be blown away in a wind than shield gas from a conventional nozzle. A slight drawback is that, like SMAW (stick) welding, there may be some flux deposited over the weld bead, requiring more of a cleaning process between passes.

Flux-cored welding machines are most popular at the hobbyist level, as the machines are slightly simpler but mainly because they avoid the cost of providing shield gas, either through a rented cylinder or with the high cost of disposable cylinders.

GMAW-based 3-D printing

GMAW has also been used a low-cost method to 3-D print metal objects. Various open source 3-D printers have been developed to use GMAW. Such components fabricated from aluminum compete with more traditionally manufactured components on mechanical strength. By forming a bad weld on the first layer, GMAW 3-D printed parts can be removed from the substrate with a hammer.

Operation

For most of its applications gas metal arc welding is a fairly simple welding process to learn requiring no more than a week or two to master basic welding technique. Even when welding is performed by well-trained operators weld quality can fluctuate since it depends on a number of external factors. All GMAW is dangerous, though perhaps less so than some other welding methods, such as shielded metal arc welding.

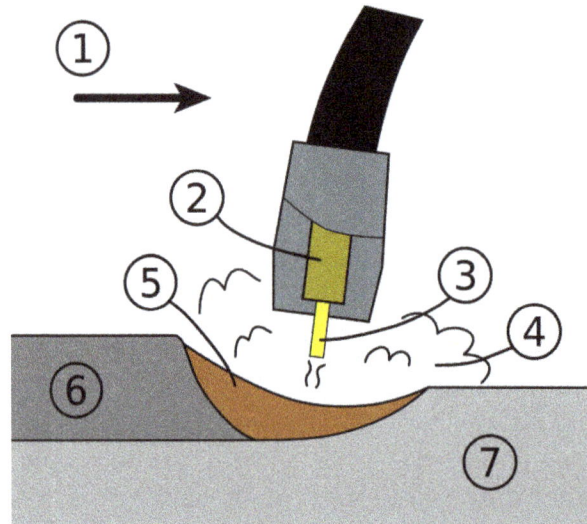

GMAW weld area. (1) Direction of travel, (2) Contact tube, (3) Electrode, (4) Shielding gas, (5) Molten weld metal, (6) Solidified weld metal, (7) Workpiece.

Technique

The basic technique for GMAW is quite simple, since the electrode is fed automatically through the torch (head of tip). By contrast, in gas tungsten arc welding, the welder must handle a welding torch in one hand and a separate filler wire in the other, and in shielded metal arc welding, the operator must frequently chip off slag and change welding electrodes. GMAW requires only that the operator guide the welding gun with proper position and orientation along the area being welded. Keeping a consistent contact tip-to-work distance (the *stick out* distance) is important, because a long stickout distance can cause the electrode to overheat and also wastes shielding gas. Stickout distance varies for different GMAW weld processes and applications. The orientation of the gun is also important—it should be held so as to bisect the angle between the workpieces; that is, at 45 degrees for a fillet weld and 90 degrees for welding a flat surface. The travel angle, or lead angle, is the angle of the torch with respect to the direction of travel, and it should generally remain approximately vertical. However, the desirable angle changes somewhat depending on the type of shielding gas used—with pure inert gases, the bottom of the torch is often slightly in front of the upper section, while the opposite is true when the welding atmosphere is carbon dioxide.

Quality

Two of the most prevalent quality problems in GMAW are dross and porosity. If not controlled, they can lead to weaker, less ductile welds. Dross is an especially common problem in aluminium GMAW welds, normally coming from particles of aluminium oxide or aluminum nitride present in the electrode or base materials. Electrodes and workpieces must be brushed with a wire brush or chemically treated to remove oxides on the surface. Any oxygen in contact with the weld pool, whether from the atmosphere or the shielding gas, causes dross as well. As a result, sufficient flow of inert shielding gases is necessary, and welding in volatile air should be avoided.

In GMAW the primary cause of porosity is gas entrapment in the weld pool, which occurs when the metal solidifies before the gas escapes. The gas can come from impurities in the shielding gas or on the workpiece, as well as from an excessively long or violent arc. Generally, the amount of gas entrapped is directly related to the cooling rate of the weld pool. Because of its higher thermal conductivity, aluminum welds are especially susceptible to greater cooling rates and thus additional porosity. To reduce it, the workpiece and electrode should be clean, the welding speed diminished and the current set high enough to provide sufficient heat input and stable metal transfer but low enough that the arc remains steady. Preheating can also help reduce the cooling rate in some cases by reducing the temperature gradient between the weld area and the base material.

Safety

Gas metal arc welding can be dangerous if proper precautions are not taken. Since GMAW employs an electric arc, welders wear protective clothing, including heavy leather gloves and protective long sleeve jackets, to avoid exposure to extreme heat and flames. In addition, the brightness of the electric arc is a source of the condition known as arc eye, an inflammation of the cornea caused by ultraviolet light and, in prolonged exposure, possible burning of the retina in the eye. Conventional welding helmets contain dark face plates to prevent this exposure. Newer helmet designs feature a liquid crystal-type face plate that self-darken upon exposure to high amounts of UV light. Transparent welding curtains, made of a polyvinyl chloride plastic film, are often used to shield nearby workers and bystanders from exposure to the UV light from the electric arc.

Welders are also often exposed to dangerous gases and particulate matter. GMAW produces smoke containing particles of various types of oxides, and the size of the particles in question tends to influence the toxicity of the fumes, with smaller particles presenting a greater danger. Additionally, carbon dioxide and ozone gases can prove dangerous if ventilation is inadequate. Furthermore, because the use of compressed gases in GMAW pose an explosion and fire risk, some common precautions include limiting the amount of oxygen in the air and keeping combustible materials away from the workplace.

Metal Transfer Modes

The three transfer modes in GMAW are globular, short-circuiting, and spray. There are a few recognized variations of these three transfer modes including modified short-circuiting and pulsed-spray.

Globular

GMAW with globular metal transfer is considered the least desirable of the three major GMAW variations, because of its tendency to produce high heat, a poor weld surface, and spatter. The method was originally developed as a cost efficient way to weld steel using GMAW, because this variation uses carbon dioxide, a less expensive shielding gas than argon. Adding to its economic advantage was its high deposition rate, allowing welding speeds of up to 110 mm/s (250 in/min). As the weld is made, a ball of molten metal from the electrode tends to build up on the end of the electrode, often in irregular shapes with a larger diameter than the electrode itself. When the droplet finally detaches either by gravity or short circuiting, it falls to the workpiece, leaving an uneven surface and often causing spatter. As a result of the large molten droplet, the process is generally limited to flat and horizontal welding positions, requires thicker workpieces, and results in a larger weld pool.

Short-circuiting

Further developments in welding steel with GMAW led to a variation known as short-circuit transfer (SCT) or short-arc GMAW, in which the current is lower than for the globular method. As a result of the lower current, the heat input for the short-arc variation is considerably reduced, making it possible to weld thinner materials while decreasing the amount of distortion and residual stress in the weld area. As in globular welding, molten droplets form on the tip of the electrode, but instead of dropping to the weld pool, they bridge the gap between the electrode and the weld pool as a result of the lower wire feed rate. This causes a short circuit and extinguishes the arc, but it is quickly reignited after the surface tension of the weld pool pulls the molten metal bead off the electrode tip. This process is repeated about 100 times per second, making the arc appear constant to the human eye. This type of metal transfer provides better weld quality and less spatter than the globular variation, and allows for welding in all positions, albeit with slower deposition of weld material. Setting the weld process parameters (volts, amps and wire feed rate) within a relatively narrow band is critical to maintaining a stable arc: generally between 100 and 200 amperes at 17 to 22 volts for most applications. Also, using short-arc transfer can result in lack of fusion and insufficient penetration when welding thicker materials, due to the lower arc energy and rapidly freezing weld pool. Like the globular variation, it can only be used on ferrous metals.

Spray

Spray transfer GMAW was the first metal transfer method used in GMAW, and well-suited to welding aluminium and stainless steel while employing an inert shielding gas. In this GMAW process, the weld electrode metal is rapidly passed along the stable electric arc from the electrode to the workpiece, essentially eliminating spatter and resulting in a high-quality weld finish. As the current and voltage increases beyond the range of short circuit transfer the weld electrode metal transfer transitions from larger globules through small droplets to a vaporized stream at the highest energies. Since this vaporized spray transfer variation of the GMAW weld process requires higher voltage and current than short circuit transfer, and as a result of the higher heat input and larger weld pool area (for a given weld electrode diameter), it is generally used only on workpieces of thicknesses above about 6.4 mm (0.25 in).

Also, because of the large weld pool, it is often limited to flat and horizontal welding positions and sometimes also used for vertical-down welds. It is generally not practical for root pass welds. When a smaller electrode is used in conjunction with lower heat input, its versatility increases. The maximum deposition rate for spray arc GMAW is relatively high—about 60 mm/s (150 in/min).

Pulsed-spray

A variation of the spray transfer mode, pulse-spray is based on the principles of spray transfer but uses a pulsing current to melt the filler wire and allow one small molten droplet to fall with each pulse. The pulses allow the average current to be lower, decreasing the overall heat input and thereby decreasing the size of the weld pool and heat-affected zone while making it possible to weld thin workpieces. The pulse provides a stable arc and no spatter, since no short-circuiting takes place. This also makes the process suitable for nearly all metals, and thicker electrode wire can be used as well. The smaller weld pool gives the variation greater versatility, making it possible to weld in all positions. In comparison with short arc GMAW, this method has a somewhat slower maximum speed (85 mm/s or 200 in/min) and the process also requires that the shielding gas be primarily argon with a low carbon dioxide concentration. Additionally, it requires a special power source capable of providing current pulses with a frequency between 30 and 400 pulses per second. However, the method has gained popularity, since it requires lower heat input and can be used to weld thin workpieces, as well as nonferrous materials.

Flux-cored Arc Welding

FCAW wire feeder

Flux-cored arc welding (FCAW or FCA) is a semi-automatic or automatic arc welding process. FCAW requires a continuously-fed consumable tubular electrode containing a flux and a constant-voltage or, less commonly, a constant-current welding power supply. An externally supplied shielding gas is sometimes used, but often the flux itself is relied upon to generate the necessary protection from the atmosphere, producing both gaseous protection and liquid slag protecting the weld. The process is widely used in construction because of its high welding speed and portability.

FCAW was first developed in the early 1950s as an alternative to shielded metal arc welding (SMAW). The advantage of FCAW over SMAW is that the use of the stick electrodes used in SMAW is unnecessary. This helped FCAW to overcome many of the restrictions associated with SMAW.

Types

One type of FCAW requires no shielding gas. This is made possible by the flux core in the tubular consumable electrode. However, this core contains more than just flux, it also contains various ingredients that when exposed to the high temperatures of welding generate a shielding gas for protecting the arc. This type of FCAW is attractive because it is portable and generally has good penetration into the base metal. Also, windy conditions need not be considered. Some disadvantages are that this process can produce excessive, noxious smoke; As with all welding processes, the proper electrode must be chosen to obtain the required mechanical properties. Operator skill is a major factor as improper electrode manipulation or machine setup can cause porosity.

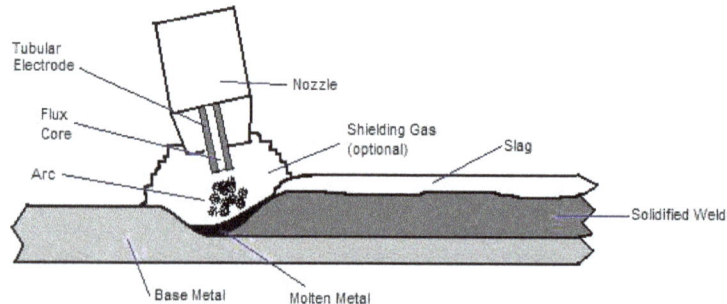

A drawing of FCAW at the weld point

Another type of FCAW uses a shielding gas that must be supplied by an external supply. This is known informally as "dual shield" welding. This type of FCAW was developed primarily for welding structural steels. In fact, since it uses both a flux-cored electrode and an external shielding gas, one might say that it is a combination of gas metal (GMAW) and flux-cored arc welding (FCAW). This particular style of FCAW is preferable for welding thicker and out-of-position metals. The slag created by the flux is also easy to remove. The main advantages of this process is that in a closed shop environment, it generally produces welds of better and more consistent mechanical properties, with fewer weld defects than either the SMAW or GMAW processes. In practice it also allows a higher production rate, since the operator does not need to stop periodically to fetch a new electrode, as is the case in SMAW. However, like GMAW, it cannot be used in a windy environment as the loss of the shielding gas from air flow will produce porosity in the weld.

Process Variables

- Wire feed speed (and current)
- Arc voltage
- Electrode extension
- Travel speed and angle
- Electrode angles
- Electrode wire type
- Shielding gas composition (if required)
- Reverse polarity (Electrode Positive) is used for FCAW Gas-Shielded wire, Straight polarity (Electrode Negative) is used for self shielded FCAW

Advantages and Applications

- FCAW may be an "all-position" process with the right filler metals (the consumable electrode)

- No shielding gas needed with some wires making it suitable for outdoor welding and/or windy conditions

- A high-deposition rate process (speed at which the filler metal is applied) in the 1G/1F/2F

- Some "high-speed" (e.g., automotive) applications

- As compared to SMAW and GTAW, there is less skill required for operators.

- Less precleaning of metal required

- Metallurgical benefits from the flux such as the weld metal being protected initially from external factors until the slag is chipped away

- Porosity chances very low

Used on the following alloys:

- Mild and low alloy steels

- Stainless steels

- Some high nickel alloys

- Some wearfacing/surfacing alloys

Disadvantages

Of course, all of the usual issues that occur in welding can occur in FCAW such as incomplete fusion between base metals, slag inclusion (non-metallic inclusions), and cracks in the welds. But there are a few concerns that come up with FCAW that are worth taking special note of:

- Melted contact tip – when the contact tip actually contacts the base metal, fusing the two and melting the hole on the end

- Irregular wire feed – typically a mechanical problem

- Porosity – the gases (specifically those from the flux-core) don't escape the welded area before the metal hardens, leaving holes in the welded metal

- More costly filler material/wire as compared to GMAW

- The equipment is less mobile and more costly as compared to SMAW or GTAW.

- The amount of smoke generated can far exceed that of SMAW, GMAW, or GTAW.

- Changing filler metals requires changing an entire spool. This can be slow and difficult as compared to changing filler metal for SMAW or GTAW.

- Creates more fumes than SMAW.

Submerged Arc Welding

Submerged arc welding (SAW) is a common arc welding process. The first patent on the submerged-arc welding (SAW) process was taken out in 1935 and covered an electric arc beneath a bed of granulated flux. Originally developed and patented by Jones, Kennedy and Rothermund, the process requires a continuously fed consumable solid or tubular (metal cored) electrode. The molten weld and the arc zone are protected from atmospheric contamination by being "submerged" under a blanket of granular fusible flux consisting of lime, silica, manganese oxide, calcium fluoride, and other compounds. When molten, the flux becomes conductive, and provides a current path between the electrode and the work. This thick layer of flux completely covers the molten metal thus preventing spatter and sparks as well as suppressing the intense ultraviolet radiation and fumes that are a part of the shielded metal arc welding (SMAW) process.

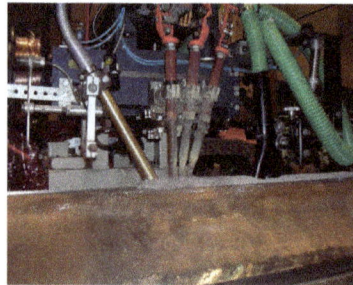

Submerged arc welding. The welding head moves from right to left. The flux powder is supplied by the hopper on the left hand side, then follow three filler wire guns and finally a vacuum cleaner.

A submerged arc welder used for training

Close-up view of the control panel

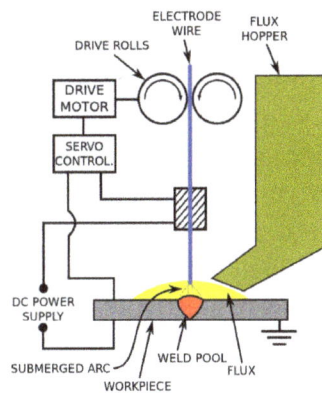

A schematic of submerged arc welding

Pieces of slag from Submerged arc welding

SAW is normally operated in the automatic or mechanized mode, however, semi-automatic (hand-held) SAW guns with pressurized or gravity flux feed delivery are available. The process is normally limited to the flat or horizontal-fillet welding positions (although horizontal groove position welds have been done with a special arrangement to support the flux). Deposition rates approaching 45 kg/h (100 lb/h) have been reported — this compares to ~5 kg/h (10 lb/h) (max) for shielded metal arc welding. Although currents ranging from 300 to 2000 A are commonly utilized, currents of up to 5000 A have also been used (multiple arcs).

Single or multiple (2 to 5) electrode wire variations of the process exist. SAW strip-cladding utilizes a flat strip electrode (e.g. 60 mm wide x 0.5 mm thick). DC or AC power can be used, and combinations of DC and AC are common on multiple electrode systems. Constant voltage welding power supplies are most commonly used; however, constant current systems in combination with a voltage sensing wire-feeder are available.

Features

Welding Head

It feeds flux and filler metal to the welding joint. Electrode (filler metal) gets energized here.

Flux Hopper

It stores the flux and controls the rate of flux deposition on the welding joint.

Flux

The granulated flux shields and thus protects molten weld from atmospheric contamination. The flux cleans weld metal and can modify its chemical composition also. The flux is granulated to a definite size. It may be of fused, bonded or mechanically mixed type. The flux may consist of fluorides of calcium and oxides of calcium, magnesium, silicon, aluminium and manganese. Alloying elements may be added as per requirements. Substances evolving large amount of gases during welding are never mixed with the flux. Flux with fine and coarse particle sizes are recommended for welding heavier and smaller thickness respectively.

Electrode

SAW filler material usually is a standard wire as well as other special forms. This wire normally has a thickness of 1.6 mm to 6 mm (1/16 in. to 1/4 in.). In certain circumstances, twisted wire can be used to give the arc an oscillating movement. This helps fuse the toe of the weld to the base metal. The electrode composition depends upon the material being welded. Alloying elements may be added in the electrodes. Electrodes are available to weld mild steels, high carbon steels, low and special alloy steels, stainless steel and some of the nonferrous of copper and nickel. Electrodes are generally copper coated to prevent rusting and to increase their electrical conductivity. Electrodes are available in straight lengths and coils. Their diameters may be 1.6, 2.0, 2.4, 3, 4.0, 4.8, and 6.4 mm. The approximate value of currents to weld with 1.6, 3.2 and 6.4 mm diameter electrodes are 150–350, 250–800 and 650–1350 Amps respectively.

Welding Operation

The flux starts depositing on the joint to be welded. Since the flux when cold is non-conductor of electricity, the arc may be struck either by touching the electrode with the work piece or by placing steel wool between electrode and job before switching on the welding current or by using a high frequency unit. In all cases the arc is struck under a cover of flux. Flux otherwise is an insulator but once it melts due to heat of the arc, it becomes highly conductive and hence the current flow is maintained between the electrode and the workpiece through the molten flux. The upper portion of the flux, in contact with atmosphere, which is visible remains granular (unchanged) and can be reused. The lower, melted flux becomes slag, which is waste material and must be removed after welding.

The electrode at a predetermined speed is continuously fed to the joint to be welded. In semi-automatic welding sets the welding head is moved manually along the joint. In automatic welding a separate drive moves either the welding head over the stationary job or the job moves/rotates under the stationary welding head.

The arc length is kept constant by using the principle of a self-adjusting arc. If the arc length decreases, arc voltage will increase, arc current and therefore burn-off rate will increase thereby causing the arc to lengthen. The reverse occurs if the arc length increases more than the normal.

A backing plate of steel or copper may be used to control penetration and to support large amounts of molten metal associated with the process.

Key SAW Process Variables

- Wire feed speed (main factor in welding current control)
- Arc voltage
- Travel speed
- Electrode stick-out (ESO) or contact tip to work (CTTW)
- Polarity and current type (AC or DC) and variable balance AC current

Material Applications

- Carbon steels (structural and vessel construction)
- Low alloy steels
- Stainless steels
- Nickel-based alloys
- Surfacing applications (wear-facing, build-up, and corrosion resistant overlay of steels)

Advantages

- High deposition rates (over 45 kg/h (100 lb/h) have been reported).
- High operating factors in mechanized applications.
- Deep weld penetration.
- Sound welds are readily made (with good process design and control).
- High speed welding of thin sheet steels up to 5 m/min (16 ft/min) is possible.
- Minimal welding fume or arc light is emitted.
- Practically no edge preparation is necessary depending on joint configuration and required penetration.
- The process is suitable for both indoor and outdoor works.
- Welds produced are sound, uniform, ductile, corrosion resistant and have good impact value.
- Single pass welds can be made in thick plates with normal equipment.
- The arc is always covered under a blanket of flux, thus there is no chance of spatter of weld.
- 50% to 90% of the flux is recoverable, recycled and reused.

Limitations

- Limited to ferrous (steel or stainless steels) and some nickel-based alloys.

- Normally limited to the 1F, 1G, and 2F positions.

- Normally limited to long straight seams or rotated pipes or vessels.

- Requires relatively troublesome flux handling systems.

- Flux and slag residue can present a health and safety concern.

- Requires inter-pass and post weld slag removal.

Electroslag Welding

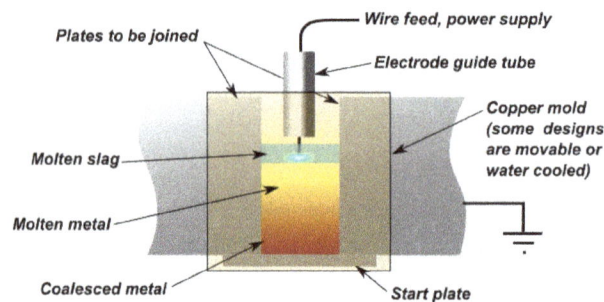

Electroslag welding (ESW) is a highly productive, single pass welding process for thick (greater than 25 mm up to about 300 mm) materials in a vertical or close to vertical position. (ESW) is similar to electrogas welding, but the main difference is the arc starts in a different location. An electric arc is initially struck by wire that is fed into the desired weld location and then flux is added. Additional flux is added until the molten slag, reaching the tip of the electrode, extinguishes the arc. The wire is then continually fed through a consumable guide tube (can oscillate if desired) into the surfaces of the metal workpieces and the filler metal are then melted using the electrical resistance of the molten slag to cause coalescence. The wire and tube then move up along the workpiece while a copper retaining shoe that was put into place before starting (can be water-cooled if desired) is used to keep the weld between the plates that are being welded. Electroslag welding is used mainly to join low carbon steel plates and/or sections that are very thick. It can also be used on structural steel if certain precautions are observed. This process uses a direct current (DC) voltage usually ranging from about 600A and 40-50V, higher currents are needed for thicker materials. Because the arc is extinguished, this is not an arc process.

History

The process was patented by Robert K Hopkins in the United States in February 1940 (patent 2191481) and developed and refined at the Paton Institute, Kiev, USSR during the 1940s. The Paton method was released to the west at the Bruxelles Trade Fair of 1950. The first widespread use in the U.S. was in 1959, by General Motors Electromotive Division, Chicago, for the fabrication of traction motor frames. In 1968 Hobart Brothers of Troy, Ohio, released a range of machines for use in the shipbuilding, bridge construction and large structural fabrication industries. Between the late 1960s and late 1980s, it is estimated that in California alone over a million stiffeners were welded with the electroslag welding process. Two of the tallest buildings in California were welded, using the electroslag welding process - The Bank of America building in San Francisco, and the twin tower Security Pacific buildings in Los Angeles. The Northridge earthquake and the Loma Prieta earthquakes provided a "real world" test to compare all of the welding processes. The Structural Steel welding industry is well aware that, over one billion dollars in crack repairs were needed, after the Northridge earthquake, to repair weld cracks propagated in welds made with the gasless flux cored wire process. Not one failure or one crack propagation was initiated in any of the hundreds-of-thousands of welds made on continuity plates welded with the Electroslag welding process. The History Of Electroslag Welding For High Rise Building And Bridges

However the Federal Highway Administration (FHWA) monitored the new process and found that electroslag welding, because of the very large amounts of confined heat used, produced a coarse-grained and brittle weld and in 1977 banned the use of the process for many applications. The FHWA commissioned research from universities and industry and Narrow Gap Improved Electro Slag Welding (NGI-ESW) was developed as a replacement. The FHWA moratorium was rescinded in 2000.

Benefits

Benefits of the process include its high metal deposition rates—it can lay metal at a rate between 15 and 20 kg per hour (35 and 45 lb/h) per electrode—and its ability to weld thick materials. Many welding processes require more than one pass for welding thick workpieces, but often a single pass is sufficient for electroslag welding. The process is also very efficient, since joint preparation and materials handling are minimized while filler metal utilization is high. The process is also safe and clean, with no arc flash and low weld splatter or distortion. Electroslag welding easily lends itself to mechanization, thus reducing the requirement for skilled manual welders.

One electrode is commonly used to make welds on materials with a thickness of 25 to 75 mm (1 to 3 in), and thicker pieces generally require more electrodes. The maximum workpiece thickness that has ever been successfully welded was a 0.91 m (36 in) piece that required the simultaneous use of six electrodes to complete.

Plasma Arc Welding

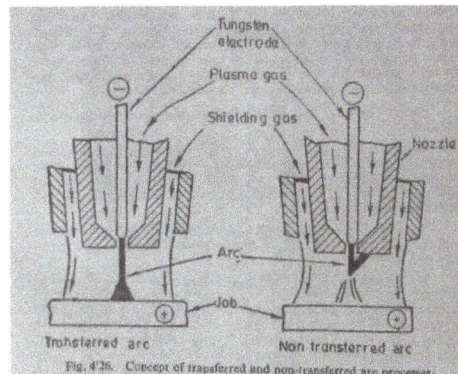

Plasma arc types.

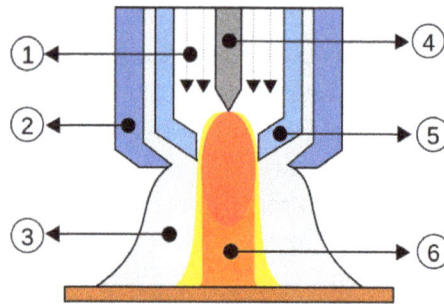

1. Gas plasma, 2. Nozzle protection, 3. Shield Gas, 4. Electrode, 5. Nozzle constriction, 6. Electric arc

Plasma arc welding (PAW) is an arc welding process similar to gas tungsten arc welding (GTAW). The electric arc is formed between an electrode (which is usually but not always made of sintered tungsten) and the workpiece. The key difference from GTAW is that in PAW, by positioning the electrode within the body of the torch, the plasma arc can be separated from the shielding gas envelope. The plasma is then forced through a fine-bore copper nozzle which constricts the arc and the plasma exits the orifice at high velocities (approaching the speed of sound) and a temperature approaching 28,000 °C (50,000 °F) or higher.

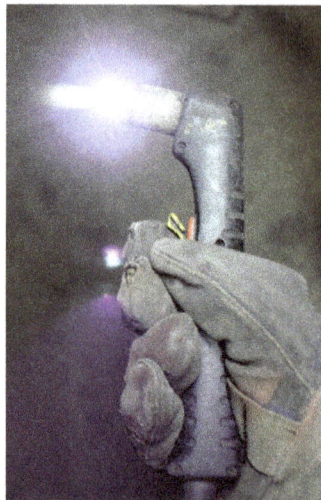

Plasma cutting torch.

Metal cutting by plasma arc.

Just as oxy-fuel torches can be used for either welding or cutting, so too can plasma torches, which can achieve plasma arc welding or plasma cutting.

Arc plasma is the temporary state of a gas. The gas gets ionized after passage of electric current through it and it becomes a conductor of electricity. In ionized state atoms break into electrons (−) and cations (+) and the system contains a mixture of ions, electrons and highly excited atoms. The degree of ionization may be between 1% and greater than 100% i.e.; double and triple degrees of ionization. Such states exist as more electrons are pulled from their orbits.

The energy of the plasma jet and thus the temperature is dependent upon the electrical power employed to create arc plasma. A typical value of temperature obtained in a plasma jet torch may be of the order of 28000 °C(50000 °F) against about 5500 °C (10000 °F) in ordinary electric welding arc. Actually all welding arcs are (partially ionized) plasmas, but the one in plasma arc welding is a constricted arc plasma.

Concept

Plasma arc welding is an arc welding process wherein coalescence is produced by the heat obtained from a constricted arc setup between a tungsten/alloy tungsten electrode and the water-cooled (constricting) nozzle (non-transferred arc) or between a tungsten/alloy tungsten electrode and the job (transferred arc). The process employs two inert gases, one forms the arc plasma and the second shields the arc plasma. Filler metal may or may not be added.

History

The plasma arc welding and cutting process was invented by Robert M. Gage in 1953 and patented in 1957. The process was unique in that it could achieve precision cutting and welding on both thin and thick metals. It was also capable of spray coating hardening metals onto other metals. One example was the spray coating of the turbine blades of the moon bound Saturn rocket.

Principle of Operation

Plasma arc welding is a constricted arc process. The arc is constricted with the help of a water-cooled small diameter nozzle which squeezes the arc, increases its pressure, temperature and heat intensely and thus improves arc stability, arc shape and heat transfer characteristics. Plasma arc welding processes can be divided into two basic types:

Non-transferred arc process

The arc is formed between the electrode(-) and the water cooled constricting nozzle(+). Arc plasma comes out of the nozzle as a flame. The arc is independent of the work piece and the work piece does not form a part of the electrical circuit. Just like an arc flame (as in atomic hydrogen welding), it can be moved from one place to another and can be better controlled. The non transferred plasma arc possesses comparatively less energy density as compared to a transferred arc plasma and it is employed for welding and in applications involving ceramics or metal plating (spraying). High density metal coatings can be produced by this process. A non-transferred arc is initiated by using a high frequency unit in the circuit.

Transferred arc process

> The arc is formed between the electrode(-) and the work piece(+). In other words, arc is transferred from the electrode to the work piece. A transferred arc possesses high energy density and plasma jet velocity. For this reason it is employed to cut and melt metals. Besides carbon steels this process can cut stainless steel and nonferrous metals where an oxyacetylene torch does not succeed. Transferred arc can also be used for welding at high arc travel speeds. For initiating a transferred arc, a current limiting resistor is put in the circuit, which permits a flow of about 50 amps, between the nozzle and electrode and a pilot arc is established between the electrode and the nozzle. As the pilot arc touches the job main current starts flowing between electrode and job, thus igniting the transferred arc. The pilot arc initiating unit gets disconnected and pilot arc extinguishes as soon as the arc between the electrode and the job is started. The temperature of a constricted plasma arc may be of the order of 8000 - 25000°C.

Equipment

The equipment needed in plasma arc welding along with their functions are as follows:

Power Supply

A direct current power source (generator or rectifier) having drooping characteristics and open circuit voltage of 70 volts or above is suitable for plasma arc welding. Rectifiers are generally preferred over DC generators. Working with helium as an inert gas needs open circuit voltage above 70 volts. This higher voltage can be obtained by series operation of two power sources; or the arc can be initiated with argon at normal open circuit voltage and then helium can be switched on.

Typical welding parameters for plasma arc welding are as follows:

Current 50 to 350 amps, voltage 27 to 31 volts, gas flow rates 2 to 40 liters/minute (lower range for orifice gas and higher range for outer shielding gas), direct current electrode negative (DCEN) is normally employed for plasma arc welding except for the welding of aluminum in which cases water cooled electrode is preferable for reverse polarity welding, i.e. direct current electrode positive (DCEP).

High Frequency Generator and Current Limiting Resistors

A high frequency generator and current limiting resistors are used for arc ignition. The arc starting system may be separate or built into the system.

Plasma Torch

It is either transferred arc or non transferred arc typed. It is hand operated or mechanized. At present, almost all applications require automated system. The torch is water cooled to increase the life of the nozzle and the electrode. The size and the type of nozzle tip are selected depending upon the metal to be welded, weld shapes and desired penetration depth.

Shielding Gases

Two inert gases or gas mixtures are employed. The orifice gas at lower pressure and flow rate forms the plasma arc. The pressure of the orifice gas is intentionally kept low to avoid weld metal turbulence, but this low pressure is not able to provide proper shielding of the weld pool. To have suitable shielding protection same or another inert gas is sent through the outer shielding ring of the torch at comparatively higher flow rates. Most of the materials can be welded with argon, helium, argon+hydrogen and argon+helium, as inert gases or gas mixtures. Argon is very commonly used. Helium is preferred where a broad heat input pattern and flatter cover pass is desired without key hole mode weld. A mixture of argon and hydrogen supplies heat energy higher than when only argon is used and thus permits keyhole mode welds in nickel base alloys, copper base alloys and stainless steels.

For cutting purposes a mixture of argon and hydrogen (10-30%) or that of nitrogen may be used. Hydrogen, because of its dissociation into atomic form and thereafter recombination generates temperatures above those attained by using argon or helium alone. In addition, hydrogen provides a reducing atmosphere, which helps in preventing oxidation of the weld and its vicinity. (Care must be taken, as hydrogen diffusing into the metal can lead to embrittlement in some metals and steels.)

Voltage Control

Voltage control is required in contour welding. In normal key hole welding a variation in arc length up to 1.5 mm does not affect weld bead penetration or bead shape to any significant extent and thus a voltage control is not considered essential.

Current and Gas Decay Control

It is necessary to close the key hole properly while terminating the weld in the structure.

Fixture

It is required to avoid atmospheric contamination of the molten metal under bead.

Process Description

Technique of work piece cleaning and filler metal addition is similar to that in TIG welding. Filler metal is added at the leading edge of the weld pool. Filler metal is not required in making root pass weld.

Type of Joints: For welding work piece up to 25 mm thick, joints like square butt, J or V are employed. Plasma welding is used to make both key hole and non-key hole types of welds.

Making a non-key hole weld: The process can make non key hole welds on work pieces having thickness 2.4 mm and under.

Making a keyhole welds: An outstanding characteristics of plasma arc welding, owing to exceptional penetrating power of plasma jet, is its ability to produce keyhole welds in work piece having thickness from 2.5 mm to 25 mm. A keyhole effect is achieved through right selection of current, nozzle orifice diameter and travel speed, which create a forceful plasma jet to penetrate completely through the work piece. Plasma jet in no case should expel the molten metal from the joint. The major advantages of keyhole technique are the ability to penetrate rapidly through relatively thick root sections and to produce a uniform under bead without mechanical backing. Also, the ratio of the depth of penetration to the width of the weld is much higher, resulting narrower weld and heat-affected zone. As the weld progresses, base metal ahead the keyhole melts, flow around the same solidifies and forms the weld bead. Key holing aids deep penetration at faster speeds and produces high quality bead. While welding thicker pieces, in laying others than root run, and using filler metal, the force of plasma jet is reduced by suitably controlling the amount of orifice gas.

Plasma arc welding is an advancement over the GTAW process. This process uses a non-consumable tungsten electrode and an arc constricted through a fine-bore copper nozzle. PAW can be used to join all metals that are weldable with GTAW (i.e., most commercial metals and alloys). Difficult-to-weld in metals by PAW include bronze, cast iron, lead and magnesium. Several basic PAW process variations are possible by varying the current, plasma gas flow rate, and the orifice diameter, including:

- Micro-plasma (< 15 Amperes)

- Melt-in mode (15–100 Amperes)

- Keyhole mode (>100 Amperes)

- Plasma arc welding has a greater energy concentration as compared to GTAW.

- A deep, narrow penetration is achievable, with a maximum depth of 12 to 18 mm (0.47 to 0.71 in) depending on the material.

- Greater arc stability allows a much longer arc length (stand-off), and much greater tolerance to arc length changes.

- PAW requires relatively expensive and complex equipment as compared to GTAW; proper torch maintenance is critical

- Welding procedures tend to be more complex and less tolerant to variations in fit-up, etc.

- Operator skill required is slightly greater than for GTAW.

- Orifice replacement is necessary.

Process Variables

Gases

At least two separate (and possibly three) flows of gas are used in PAW:

- Plasma gas – flows through the orifice and becomes ionized.

- Shielding gas – flows through the outer nozzle and shields the molten weld from the atmosphere

- Back-purge and trailing gas – required for certain materials and applications.

These gases can all be same, or of differing composition.

Key Process Variables

- Current Type and Polarity

- DCEN from a CC source is standard

- AC square-wave is common on aluminum and magnesium

- Welding current and pulsing - Current can vary from 0.5 A to 1200 A; Current can be constant or pulsed at frequencies up to 20 kHz

- Gas flow rate (This critical variable must be carefully controlled based upon the current, orifice diameter and shape, gas mixture, and the base material and thickness.)

Other Plasma Arc Processes

Depending upon the design of the torch (e.g., orifice diameter), electrode design, gas type and velocities, and the current levels, several variations of the plasma process are achievable, including:

- Plasma arc cutting (PAC)

- Plasma arc gouging

- Plasma arc surfacing

- Plasma arc spraying

Plasma Arc Cutting

When used for cutting, the plasma gas flow is increased so that the deeply penetrating plasma jet cuts through the material and molten material is removed as cutting dross. PAC differs from oxy-fuel cutting in that the plasma process operates by using the arc to melt the metal whereas in the oxy-fuel process, the oxygen oxidizes the metal and the heat from the exothermic reaction melts the metal. Unlike oxy-fuel cutting, the PAC process can be applied to cutting metals which form refractory oxides such as stainless steel, cast iron, aluminum, and other non-ferrous alloys. Since PAC was introduced by Praxair Inc. at the American Welding Society show in 1954, many process refinements, gas developments, and equipment improvements have occurred.

Spot Welding

Resistance spot welding (RSW) is a process in which contacting metal surfaces are joined by the heat obtained from resistance to electric current.

Work-pieces are held together under pressure exerted by electrodes. Typically the sheets are in the 0.5 to 3 mm (0.020 to 0.118 in) thickness range. The process uses two shaped copper alloy electrodes to concentrate welding current into a small "spot" and to simultaneously clamp the sheets together. Forcing a large current through the spot will melt the metal and form the weld. The attractive feature of spot welding is that a lot of energy can be delivered to the spot in a very short time (approximately 10–100 milliseconds). That permits the welding to occur without excessive heating of the remainder of the sheet.

A spot welder

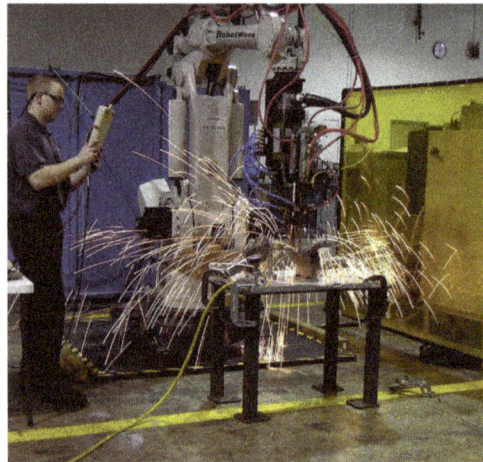

A spot welding robot

The amount of heat (energy) delivered to the spot is determined by the resistance between the electrodes and the magnitude and duration of the current. The amount of energy is chosen to match the sheet's material properties, its thickness, and type of electrodes. Applying too little energy will not melt the metal or will make a poor weld. Applying too much energy will melt too much metal, eject molten material, and make a hole rather than a weld. Another feature of spot welding is that the energy delivered to the spot can be controlled to produce reliable welds.

Projection welding is a modification of spot welding. In this process, the weld is localized by means of raised sections, or projections, on one or both of the workpieces to be joined. Heat is concentrated at the projections, which permits the welding of heavier sections or the closer spacing of welds. The projections can also serve as a means of positioning the workpieces. Projection welding is often used to weld studs, nuts, and other screw machine parts to metal plate. It is also frequently used to join crossed wires and bars. This is another high-production process, and multiple projection welds can be arranged by suitable designing and jigging.

Applications

Spot welding is typically used when welding particular types of sheet metal, welded wire mesh or wire mesh. Thicker stock is more difficult to spot weld because the heat flows into the surrounding metal more easily. Spot welding can be easily identified on many sheet metal goods, such as metal buckets. Aluminium alloys can be spot welded, but their much higher thermal conductivity and electrical conductivity requires higher welding currents. This requires larger, more powerful, and more expensive welding transformers.

BMW plant in Leipzig, Germany: Spot welding of BMW 3 series car bodies with KUKA industrial robots.

Perhaps the most common application of spot welding is in the automobile manufacturing industry, where it is used almost universally to weld the sheet metal to form a car. Spot welders can also be completely automated, and many of the industrial robots found on assembly lines are spot welders (the other major use for robots being painting).

Spot welding is also used in the orthodontist's clinic, where small-scale spot welding equipment is used when resizing metal "molar bands" used in orthodontics.

Another application is spot welding straps to nickel–cadmium or nickel–metal hydride cells to make batteries. The cells are joined by spot welding thin nickel straps to the battery terminals. Spot welding can keep the battery from getting too hot, as might happen if conventional soldering were done.

Good design practice must always allow for adequate accessibility. Connecting surfaces should be free of contaminants such as scale, oil, and dirt, to ensure quality welds. Metal thickness is generally not a factor in determining good welds.

Processing and Equipment

Spot welding involves three stages; the first of which involves the electrodes being brought to the surface of the metal and applying a slight amount of pressure. The current from the electrodes is then applied briefly after which the current is removed but the electrodes remain in place for the material to cool. Weld times range from 0.01 sec to 0.63 sec depending on the thickness of the metal, the electrode force and the diameter of the electrodes themselves.

The equipment used in the spot welding process consists of tool holders and electrodes. The tool holders function as a mechanism to hold the electrodes firmly in place and also support optional water hoses that cool the electrodes during welding. Tool holding methods include a paddle-type, light duty, universal, and regular offset. The electrodes generally are made of a low resistance alloy, usually copper, and are designed in many different shapes and sizes depending on the application needed.

The two materials being welded together are known as the workpieces and must conduct electricity. The width of the workpieces is limited by the throat length of the welding apparatus and ranges typically from 5 to 50 inches (13 to 130 cm). Workpiece thickness can range from 0.008 to 1.25 inches (0.20 to 32 mm).

After the current is removed from the workpiece, it is cooled via the coolant holes in the center of the electrodes. Both water and a brine solution may be used as coolants in spot welding mechanisms.

Tool Styles

Electrodes used in spot welding can vary greatly with different applications. Each tool style has a different purpose. Radius style electrodes are used for high heat applications, electrodes with a truncated tip for high pressure, eccentric electrodes for welding corners, offset eccentric tips for reaching into corners and small spaces, and finally offset truncated for reaching into the workpiece itself.

Effects

The spot welding process tends to harden the material, causing it to warp. This reduces the material's fatigue strength, and may stretch the material as well as anneal it. The physical effects of spot welding include internal cracking, surface cracks and a bad appearance. The chemical properties affected include the metal's internal resistance and its corrosive properties.

Electrical Notes

The basic spot welder consists of a power supply, an energy storage unit (e.g., a capacitor bank), a switch, a welding transformer, and the welding electrodes. The energy storage element allows the welder to deliver high instantaneous power levels. If the power demands are not high, then the energy storage element isn't needed. The switch causes the stored energy to be dumped into the welding transformer. The welding transformer steps down the voltage and steps up the current. An important feature of the transformer is it reduces the current level that the switch must handle. The welding electrodes are part of the transformer's secondary circuit. There is also a control box that manages the switch and may monitor the welding electrode voltage or current.

The resistance presented to the welder is complicated. There is the resistance of secondary winding, the cables, and the welding electrodes. There is also the contact resistance between the welding electrodes and the workpiece. There is the resistance of the workpieces, and the contact resistance between the workpieces.

At the beginning of the weld, the contact resistances are usually high, so most of the initial energy will be dissipated there. That heat and the clamping force will soften and smooth out the material at the electrode-material interface and make better contact (that is, lower the contact resistance). Consequently, more electrical energy will go into the workpiece and the junction resistance of the two workpieces. As electrical energy is delivered to the weld and causes the temperature to rise, the electrodes and the workpiece are conducting that heat away. The goal is to apply enough energy so that a portion of material within the spot melts without having the entire spot melt. The perimeter of the spot will conduct away a lot of heat and keep the perimeter at a lower temperature. The interior of the spot has less heat conducted away, so it melts first. If the welding current is applied too long, the entire spot melts, the material runs out or otherwise fails, and the "weld" becomes a hole.

The voltage needed for welding depends on the resistance of the material to be welded, the sheet thickness and desired size of the nugget. When welding a common combination like 1.0 + 1.0 mm sheet steel, the voltage between the electrodes is only about 1.5 V at the start of the weld but can fall as low as 1 V at the end of the weld. This decrease in voltage results from the reduction in resistance caused by the workpiece melting. The open circuit voltage from the transformer is higher than this, typically in the 5 to 22 volt range.

The resistance of the weld spot changes as it flows and liquefies. Modern welding equipment can monitor and adjust the weld in real-time to ensure a consistent weld. The equipment may seek to control different variables during the weld, such as current, voltage, power, or energy.

Welder sizes range from 5 to 500 kVA. Micro spot welders, used in a variety of industries, can go down to 1.5 kVA or less for precision welding needs.

Physics

Clamping

Welding times are often very short, which can cause problems with the electrodes—they cannot move fast enough to keep the material clamped. Welding controllers will use a double pulse to get around this problem. During the first pulse, the electrode contact may not be able to make a good weld. The first pulse will soften the metal. During the pause between the two pulses, the electrodes will come closer and make better contact.

Fields

During spot welding, the large electric current induces a large magnetic field, and the electric current and magnetic field interact with each other to produce a large magnetic force field too, which drives the melted metal to move very fast at a velocity up to 0.5 m/s. As such, the heat energy distribution in spot welding could be dramatically changed by the fast motion of the melted metal. The fast motion in spot welding can be observed with high speed photography.

Safety

It is common for a spray of molten metal droplets (sparks) to be ejected from the area of the weld during the process.

Although spot welding does not generate UV light as intensely as arc welding, eye protection is nevertheless required. Welding goggles with a 5.0 shade are recommended.

Forge Welding

Forge welding (FOW) is a solid-state welding process that joins two pieces of metal by heating them to a high temperature and then hammering them together. It may also consist of heating and forcing the metals together with presses or other means, creating enough pressure to cause plastic deformation at the weld surfaces. The process is one of the simplest methods of joining metals and has been used since ancient times. Forge welding is versatile, being able to join a host of similar and dissimilar metals. With the invention of electrical and gas welding methods during the Industrial Revolution, manual forge-welding has been largely replaced, although automated forge-welding is a common manufacturing process.

Introduction

Forge welding is a process of joining metals by heating them beyond a certain threshold and forcing them together with enough pressure to cause deformation of the weld surfaces, creating a metallic bond between the atoms of the metals. The pressure required varies, depending on the temperature, strength, and hardness of the alloy. Forge welding is the oldest welding technique, and has been used since ancient times.

Welding processes can generally be grouped into two categories: fusion and diffusion welding. Fusion welding involves localized melting of the metals at the weld interfaces, and is common in electric or gas welding techniques. This requires temperatures much higher than the melting point of the metal in order to cause localized melting before the heat can thermally conduct away from the weld, and often a filler metal is used to keep the weld from segregating. Diffusion welding consists of joining the metals without melting them, welding the surfaces together while in the solid state. In diffusion welding, the heat source is often lower than the melting point of the metal, allowing more even heat-distribution thus reducing thermal stresses at the weld. In this method a filler metal is typically not used, but the weld occurs directly between the metals at the weld interface. This includes methods such as cold welding, explosion welding, and forge welding. Unlike other diffusion methods, in forge welding the metals are heated to a high temperature before forcing them together, usually resulting in greater plasticity at the weld surfaces. This generally makes forge welding more versatile than cold-diffusion techniques, which are usually performed on soft metals like copper or aluminum. In forge welding, the entire welding areas are heated evenly. Forge welding can be used for a much wider range of harder metals and alloys, like steel and titanium.

History

The history of joining metals goes back to the Bronze age, where bronzes of different hardness were often joined by casting-in. This method consisted of placing a solid part into a molten metal contained in a mold and allowing it to solidify without actually melting both metals, such as the blade of a sword into a handle or the tang of an arrowhead into the tip. Brazing and soldering were also common during the Bronze age. The act of welding (joining two solid parts through diffusion) began with iron. The first welding process was forge welding, which started when humans learned to smelt iron from iron ore; most likely in Anatolia (Turkey) around 1800 BC. Ancient people could not create temperatures high enough to melt iron fully, so the bloomery process that was used for smelting iron produced a lump (bloom) of iron grains sintered together, small amounts of steel, slag, and other impurities, referred to as sponge iron because of its porosity. After smelting the sponge iron needed to be welded, or "wrought," into a solid block (billet) to squeeze out air pockets and excess slag. Many items made of wrought iron have been found by archeologists, that show evidence of forge welding, which date from before 1000 BC. Because iron was typically made in small amounts, any large object, such as the Delhi Pillar, needed to be forged welded out of smaller billets.

Sponge iron used to forge a Japanese katana.

Forge welding grew from a trial-and-error method, becoming more refined over the centuries. Due to the poor quality of ancient metals, it was commonly employed in making composite steels, by joining high-carbon steels, that would resist deformation but break easily, with low-carbon steels, which resist fracture but bend too easily, creating an object with greater toughness and strength than could be produced with a single alloy. This method of pattern welding first appeared around 700 BC, and was primarily used for making weapons such as swords, with the most widely known examples being from Damascus, Japanese, and Merovingian swords. This process was also common in the manufacture of tools, from wrought-iron plows with steel edges to iron chisels with steel cutting surfaces.

Materials

Many metals can be forge welded, with the most common being both high and low-carbon steels. Iron and even some hypoeutectic cast-irons can be forge welded. Some aluminum alloys can also be forge welded. Metals such as copper, bronze and brass do not forge weld readily. Although it is possible to forge weld copper-based alloys, it is often with great difficulty due to copper's tendency to absorb oxygen during the heating. Copper and its alloys are usually better joined with cold welding, explosion welding, or other pressure-welding techniques. With iron or steel, the presence of even small amounts of copper severely reduces the alloy's ability to forge weld.

Titanium alloys are commonly forge welded. Because of titanium's tendency to absorb oxygen when molten, the solid-state, diffusion bond of a forge weld is often stronger than a fusion weld in which the metal is liquefied.

Forge welding between similar materials is caused by solid-state diffusion. This results in a weld that consists of only the welded materials without any fillers or bridging materials. Forge welding between dissimilar materials is caused by the formation of a lower melting temperature eutectic between the materials. Due to this the weld is often stronger than the individual metals.

Processes

A mechanized trip hammer.

The most well-known and oldest forge-welding process is the manual-hammering method. Manual hammering is done by heating the metal to the proper temperature, coating with flux, overlapping the weld surfaces, and then striking the joint repeatedly with a hand-held hammer. The weld surfaces are usually formed for the proper joint, and then struck with a hammer to join them. The joint is often formed to allow space for the flux to flow out, by beveling or rounding the surfaces slightly, and hammered in a successively outward fashion to squeeze the flux out. The hammer blows are typically not as hard as those used for shaping, preventing the flux from being blasted out of the joint at the first blow.

When mechanical hammers were developed, forge welding could be accomplished by heating the metal, and then placing it between the mechanized hammer and the anvil. Originally powered by waterwheels, modern mechanical-hammers can also be operated by compressed air, electricity, steam, gas engines, and many other ways. Another method is forge welding with a die, whereas the pieces of metal are heated and then forced into a die which both provides the pressure for the weld and keeps the joint at the finished shape. Roll welding is another forge welding process, where the heated metals are overlapped and passed through rollers at high pressures to create the weld.

Modern forge-welding is often automated, using computers, machines, and sophisticated hydraulic-presses to produce a variety of products from a number of various alloys. For example, steel pipe is often forge-welded during the manufacturing process. Flat stock is heated and fed through specially-shaped rollers that both form the steel into a tube and simultaneously provide the pressure to weld the edges into a continuous seam. Diffusion bonding is a common method for forge welding titanium alloys in the aerospace industry. In this process the metal is heated while in a press or die. Beyond a specific critical-temperature, which varies depending on the alloy, the impurities burn out and the surfaces are forced together. Other methods include flash welding and percussion welding. These are resistance forge-welding techniques where the press or die is electrified, passing high current through the alloy to create the heat for the weld. Shielded active-gas forge-welding is a process of forge welding in an oxygen-reactive environment, to burn out oxides, using hydrogen gas and induction heating.

Temperature

The temperature required to forge weld is typically 50 to 90 percent of the melting temperature. Iron can be welded when it surpasses the critical temperature (the A_4 temperature) where its allotrope changes from gamma iron (face-centered cubic) to delta iron (body-centered cubic). Since the critical temperatures are affected by alloying agents like carbon, steel welds at a lower temperature-range than iron. As the carbon content in the steel increases, the welding temperature-range decreases in a linear fashion. Iron, different steels, and even cast-iron can be welded to each other, provided that their carbon content is close enough that the welding ranges overlap. Pure iron can be welded when nearly white hot; between 2,500 °F (1,400 °C) and 2,700 °F (1,500 °C). Steel with a carbon content of 2.0% can be welded when orangish-yellow, between 1,700 °F (900 °C) and 2,000 °F (1,100 °C). Common steel, between 0.2 and 0.8% carbon, is typically welded at a bright yellow heat.

A primary requirement for forge welding is that both weld surfaces need to be heated to the same temperature and welded before they cool too much. When steel reaches the proper temperature, it begins to weld very readily, so a thin rod or nail heated to the same temperature will tend to stick at first contact, requiring it to be bent or twisted loose. One of the simplest ways to tell if iron or steel is hot enough is to stick a magnet to it. When iron or steel cross the A_2 critical temperature, it begins to change into the allotrope called gamma iron. When this happens, the steel or iron becomes non-magnetic. In steel, the carbon begins to mix with gamma iron at the A_3 temperature, forming a solid solution called austenite. When it crosses the A_4 critical temperature, it changes into delta iron, which is magnetic. Therefore, a blacksmith can tell when the welding temperature is reached by placing a magnet in contact with the metal. When red or orange-hot, a magnet will not stick to the metal, but when the welding temperature is crossed, the magnet will again stick to it. The steel may take on a glossy or wet appearance at the welding temperature. Care must be taken to avoid overheating the metal to the point that it gives off sparks from rapid oxidation (burning), or else the weld will be poor and brittle.

Decarburization

When steel is heated to an austenizing temperature, the carbon begins to diffuse through the iron. The higher the temperature; the greater the rate of diffusion. At such high temperatures, carbon readily combines with oxygen to form carbon dioxide, so the carbon can easily diffuse out of the steel and into the surrounding air. By the end of a blacksmithing job, the steel will be of a lower carbon content than it was prior to heating. Therefore, most blacksmithing operations are done as quickly as possible to reduce decarburization, preventing the steel from becoming too soft.

To produce the right amount of hardness in the finished product, the smith generally begins with steel that has a carbon content that is higher than desired. In ancient times, forging often began with steel that had a carbon content much too high for normal use. Most ancient forge-welding began with hypereutectoid steel, containing a carbon content sometimes well above 1.0%. Hypereutectoid steels are typically too brittle to be useful in a finished product, but by the end of forging the steel typically had a high carbon-content ranging from 0.8% (eutectoid tool-steel) to 0.5% (hypoeutectoid spring-steel).

Applications

Forge welding has been used throughout its history for making most any items out of steel and iron. It has been used in everything from the manufacture of tools, farming implements, and cookware to the manufacture of fences, gates, and prison cells. In the early Industrial Revolution, it was commonly used in the manufacture of boilers and pressure vessels, until the introduction of fusion welding. It was commonly used through the Middle Ages for producing armor and weapons.

One of the most famous applications of forge welding involves the production of pattern-welded blades. During this process a smith repeatedly draws out a billet of steel, folds it back and welds it upon itself. Another application was the manufacture of shotgun barrels. Metal wire was spooled onto a mandrel, and then forged into a barrel that was thin, uniform, and strong. In some cases the forge-welded objects are acid-etched to expose the underlying pattern of metal, which is unique to each item and provides aesthetic appeal.

Despite its diversity, forge welding had many limitations. A primary limitation was the size of objects that could be forge welded. Larger objects required a bigger heat source, and size reduced the ability to manually weld it together before it cooled too much. Welding large items like steel plate or girders was typically not possible, or at least highly impractical, until the invention of fusion welding, requiring them to be riveted instead. In some cases, fusion welding produced a much stronger weld, such as in the construction of boilers.

Flux

Forge welding requires the weld surfaces to be extremely clean or the metal will not join properly, if at all. Oxides tend to form on the surface while impurities like phosphorus and sulfur tend to migrate to the surface. Often a flux is used to keep the welding surfaces from oxidizing, which would produce a poor quality weld, and to extract other impurities from the metal. The flux mixes with the oxides that form and lowers the melting temperature and the viscosity of the oxides. This enables the oxides to flow out of the joint when the two pieces are beaten together. A simple flux can be made from borax, sometimes with the addition of powdered iron-filings.

The oldest flux used for forge welding was fine silica sand. The iron or steel would be heated in a reducing environment within the coals of the forge. Devoid of oxygen, the metal forms a layer of iron-oxide called wustite on its surface. When the metal is hot enough, but below the welding temperature, the smith sprinkles some sand onto the metal. The silicon in the sand reacts with the wustite to form fayalite, which melts just below the welding temperature. This produced a very effective flux which helped to make a strong weld.

Early examples of flux used different combinations and various amounts of iron fillings, borax, sal ammoniac, balsam of copaiba, cyanide of potash, and soda phosphate. The 1920 edition of *Scientific American book of facts and formulae* indicates a frequently offered trade secret as using copperas, saltpeter, common salt, black oxide of manganese, prussiate of potash, and "nice welding sand" (silicate).

Friction Stir Welding

Close-up view of a friction stir weld tack tool.

The bulkhead and nosecone of the Orion spacecraft are joined using friction stir welding.

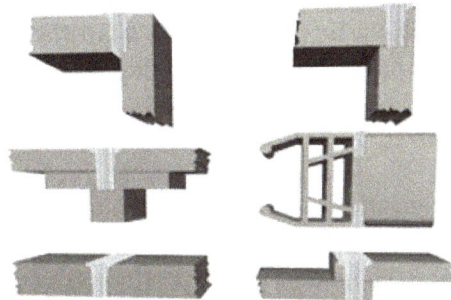

Joint designs

Friction stir welding (FSW) is a solid-state joining process that uses a non-consumable tool to join two facing workpieces without melting the workpiece material. Heat is generated by friction between the rotating tool and the workpiece material, which leads to a softened region near the FSW tool. While the tool is traversed along the joint line, it mechanically intermixes the two pieces of metal, and forges the hot and softened metal by the mechanical pressure, which is applied by the tool, much like joining clay, or dough. It is primarily used on wrought or extruded aluminium and particularly for structures which need very high weld strength.

It was invented and experimentally proven at The Welding Institute (TWI) in the UK in December 1991. TWI held patents on the process, the first being the most descriptive.

Principle of Operation

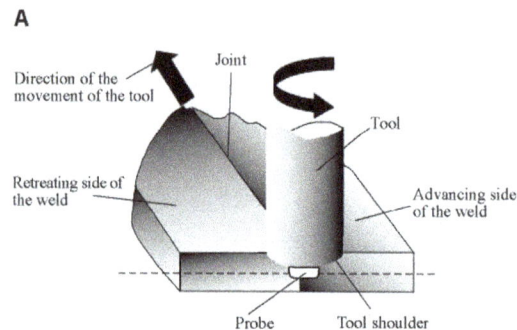

A

Schematic diagram of the FSW process: (A) Two discrete metal workpieces butted together, along with the tool (with a probe).

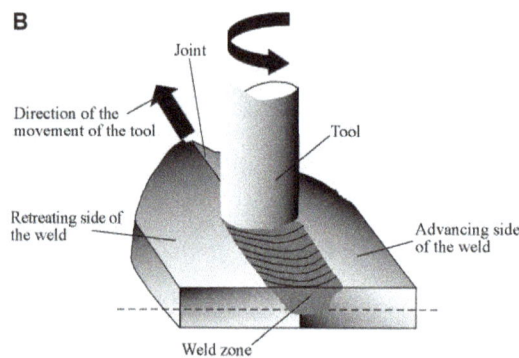

B

(B) The progress of the tool through the joint, also showing the weld zone and the region affected by the tool shoulder.

A rotating cylindrical tool with a profiled probe is fed into a butt joint between two clamped workpieces, until the shoulder, which has a larger diameter than the pin, touches the surface of the workpieces. The probe is slightly shorter than the weld depth required, with the tool shoulder riding atop the work surface. After a short dwell time, the tool is moved forward along the joint line at the pre-set welding speed.

Frictional heat is generated between the wear-resistant tool and the work pieces. This heat, along with that generated by the mechanical mixing process and the adiabatic heat within the material, cause the stirred materials to soften without melting. As the tool is moved forward, a special profile on the probe forces plasticised material from the leading face to the rear, where the high forces assist in a forged consolidation of the weld.

This process of the tool traversing along the weld line in a plasticised tubular shaft of metal results in severe solid state deformation involving dynamic recrystallization of the base material.

Microstructural Features

The solid-state nature of the FSW process, combined with its unusual tool shape and asymmetric speed profile, results in a highly characteristic microstructure. The microstructure can be broken up into the following zones:

- The *stir zone* (also nugget, dynamically recrystallised zone) is a region of heavily deformed material that roughly corresponds to the location of the pin during welding. The grains within the stir zone are roughly equiaxed and often an order of magnitude smaller than the grains in the parent material. A unique feature of the stir zone is the common occurrence of several concentric rings which has been referred to as an "onion-ring" structure. The precise origin of these rings has not been firmly established, although variations in particle number density, grain size and texture have all been suggested.

- The *flow arm zone* is on the upper surface of the weld and consists of material that is dragged by the shoulder from the retreating side of the weld, around the rear of the tool, and deposited on the advancing side.

- The *thermo-mechanically affected zone* (TMAZ) occurs on either side of the stir zone. In this region the strain and temperature are lower and the effect of welding on the microstructure is correspondingly smaller. Unlike the stir zone the microstructure is recognizably that of the parent material, albeit significantly deformed and rotated. Although the term TMAZ technically refers to the entire deformed region it is often used to describe any region not already covered by the terms stir zone and flow arm.

- The *heat-affected zone* (HAZ) is common to all welding processes. As indicated by the name, this region is subjected to a thermal cycle but is not deformed during welding. The temperatures are lower than those in the TMAZ but may still have a significant effect if the microstructure is thermally unstable. In fact, in age-hardened aluminium alloys this region commonly exhibits the poorest mechanical properties.

Advantages and Limitations

The solid-state nature of FSW leads to several advantages over fusion welding methods as problems associated with cooling from the liquid phase are avoided. Issues such as porosity, solute redistribution, solidification cracking and liquation cracking do not arise during FSW. In general, FSW has been found to produce a low concentration of defects and is very tolerant of variations in parameters and materials.

Nevertheless, FSW is associated with a number of unique defects, if it isn't done properly. Insufficient weld temperatures, due to low rotational speeds or high traverse speeds, for example, mean that the weld material is unable to accommodate the extensive deformation during welding. This may result in long, tunnel-like defects running along the weld which may occur on the surface or subsurface. Low temperatures may also limit the forging action of the tool and so reduce the continuity of the bond between the material from each side of the weld. The light contact between the material has given rise to the name "kissing-bond". This defect is particularly worrying since it is very difficult to detect using nondestructive methods such as X-ray or ultrasonic testing. If the pin is not long enough or the tool rises out of the plate then the interface at the bottom of the weld may not be disrupted and forged by the tool, resulting in a lack-of-penetration defect. This is essentially a notch in the material which can be a potential source of fatigue cracks.

A number of potential advantages of FSW over conventional fusion-welding processes have been identified:

- Good mechanical properties in the as-welded condition

- Improved safety due to the absence of toxic fumes or the spatter of molten material.

- No consumables — A threaded pin made of conventional tool steel, e.g., hardened H13, can weld over 1 km (0.62 mi) of aluminium, and no filler or gas shield is required for aluminium.

- Easily automated on simple milling machines — lower setup costs and less training.

- Can operate in all positions (horizontal, vertical, etc.), as there is no weld pool.

- Generally good weld appearance and minimal thickness under/over-matching, thus reducing the need for expensive machining after welding.

- Can use thinner materials with same joint strength.

- Low environmental impact.

- General performance and cost benefits from switching from fusion to friction.

However, some disadvantages of the process have been identified:

- Exit hole left when tool is withdrawn.

- Large down forces required with heavy-duty clamping necessary to hold the plates together.

- Less flexible than manual and arc processes (difficulties with thickness variations and non-linear welds).

- Often slower traverse rate than some fusion welding techniques, although this may be offset if fewer welding passes are required.

Important Welding Parameters

Tool Design

Advanced friction stir welding and processing tools by MegaStir shown upside down

FSW of two USIBOR 1500 high-strength steel sheets

The design of the tool is a critical factor as a good tool can improve both the quality of the weld and the maximum possible welding speed.

It is desirable that the tool material be sufficiently strong, tough, and hard wearing at the welding temperature. Further it should have a good oxidation resistance and a low thermal conductivity to minimise heat loss and thermal damage to the machinery further up the drive train. Hot-worked tool steel such as AISI H13 has proven perfectly acceptable for welding aluminium alloys within thickness ranges of 0.5 – 50 mm but more advanced tool materials are necessary for more demanding applications such as highly abrasive metal matrix composites or higher melting point materials such as steel or titanium.

Improvements in tool design have been shown to cause substantial improvements in productivity and quality. TWI has developed tools specifically designed to increase the penetration depth and thus increasing the plate thicknesses that can be successfully welded. An example is the "whorl" design that uses a tapered pin with re-entrant features or a variable pitch thread to improve the downwards flow of material. Additional designs include the Triflute and Trivex series. The Triflute design has a complex system of three tapering, threaded re-entrant flutes that appear to increase material movement around the tool. The Trivex tools use a simpler, non-cylindrical, pin and have been found to reduce the forces acting on the tool during welding.

The majority of tools have a concave shoulder profile which acts as an escape volume for the material displaced by the pin, prevents material from extruding out of the sides of the shoulder and maintains downwards pressure and hence good forging of the material behind the tool. The Triflute tool uses an alternative system with a series of concentric grooves machined into the surface which are intended to produce additional movement of material in the upper layers of the weld.

Widespread commercial applications of friction stir welding process for steels and other hard alloys such as titanium alloys will require the development of cost-effective and durable tools. Material selection, design and cost are important considerations in the search for commercially useful tools for the welding of hard materials. Work is continuing to better understand the effects of tool material's composition, structure, properties and geometry on their performance, durability and cost.

Tool Rotation and Traverse Speeds

There are two tool speeds to be considered in friction-stir welding; how fast the tool rotates and how quickly it traverses along the interface. These two parameters have considerable importance and must be chosen with care to ensure a successful and efficient welding cycle. The relationship between the rotation speed, the welding speed and the heat input during welding is complex but, in general, it can be said that increasing the rotation speed or decreasing the traverse speed will result in a hotter weld. In order to produce a successful weld it is necessary that the material surrounding the tool is hot enough to enable the extensive plastic flow required and minimize the forces acting on the tool. If the material is too cold then voids or other flaws may be present in the stir zone and in extreme cases the tool may break.

Excessively high heat input, on the other hand may be detrimental to the final properties of the weld. Theoretically, this could even result in defects due to the liquation of low-melting-point phases (similar to liquation cracking in fusion welds). These competing demands lead onto the concept of a "processing window": the range of processing parameters viz. tool rotation and traverse speed, that will produce a good quality weld. Within this window the resulting weld will have a sufficiently high heat input to ensure adequate material plasticity but not so high that the weld properties are excessively deteriorated.

Tool Tilt and Plunge Depth

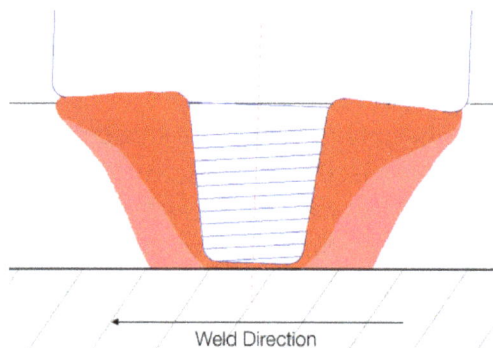

A drawing showing the plunge depth and tilt of the tool. The tool is moving to the left.

The plunge depth is defined as the depth of the lowest point of the shoulder below the surface of the welded plate and has been found to be a critical parameter for ensuring weld quality. Plunging the shoulder below the plate surface increases the pressure below the tool and helps ensure adequate forging of the material at the rear of the tool. Tilting the tool by 2–4 degrees, such that the rear of the tool is lower than the front, has been found to assist this forging process. The plunge depth needs to be correctly set, both to ensure the necessary downward pressure is achieved and to ensure that the tool fully penetrates the weld. Given the high loads required, the welding machine may deflect and so reduce the plunge depth compared to the nominal setting, which may result in flaws in the weld. On the other hand, an excessive plunge depth may result in the pin rubbing on the backing plate surface or a significant undermatch of the weld thickness compared to the base material. Variable load welders have been developed to automatically compensate for changes in the tool displacement while TWI have demonstrated a roller system that maintains the tool position above the weld plate.

Welding Forces

During welding a number of forces will act on the tool:

- A downwards force is necessary to maintain the position of the tool at or below the material surface. Some friction-stir welding machines operate under load control but in many cases the vertical position of the tool is preset and so the load will vary during welding.

- The traverse force acts parallel to the tool motion and is positive in the traverse direction. Since this force arises as a result of the resistance of the material to the motion of the tool it might be expected that this force will decrease as the temperature of the material around the tool is increased.

- The lateral force may act perpendicular to the tool traverse direction and is defined here as positive towards the advancing side of the weld.

- Torque is required to rotate the tool, the amount of which will depend on the down force and friction coefficient (sliding friction) and/or the flow strength of the material in the surrounding region (stiction).

In order to prevent tool fracture and to minimize excessive wear and tear on the tool and associated machinery, the welding cycle is modified so that the forces acting on the tool are as low as possible, and abrupt changes are avoided. In order to find the best combination of welding parameters, it is likely that a compromise must be reached, since the conditions that favour low forces (e.g. high heat input, low travel speeds) may be undesirable from the point of view of productivity and weld properties.

Flow of Material

Early work on the mode of material flow around the tool used inserts of a different alloy, which had a different contrast to the normal material when viewed through a microscope, in an effort to determine where material was moved as the tool passed. The data was interpreted as representing a form of in-situ extrusion where the tool, backing plate and cold base material form the "extrusion chamber" through which the hot, plasticised material is forced. In this model the rotation of the tool draws little or no material around the front of the probe instead the material parts in front of the pin and passes down either side. After the material has passed the probe the side pressure exerted by the "die" forces the material back together and consolidation of the join occurs as the rear of the tool shoulder passes overhead and the large down force forges the material.

More recently, an alternative theory has been advanced that advocates considerable material movement in certain locations. This theory holds that some material does rotate around the probe, for at least one rotation, and it is this material movement that produces the "onion-ring" structure in the stir zone. The researchers used a combination of thin copper strip inserts and a "frozen pin" technique, where the tool is rapidly stopped in place. They suggested that material motion occurs by two processes:

1. Material on the advancing side of a weld enters into a zone that rotates and advances with the profiled probe. This material was very highly deformed and sloughs off behind the pin

to form arc-shaped features when viewed from above (i.e. down the tool axis). It was noted that the copper entered the rotational zone around the pin, where it was broken up into fragments. These fragments were only found in the arc shaped features of material behind the tool.

2. The lighter material came from the retreating side in front of the pin and was dragged around to the rear of the tool and filled in the gaps between the arcs of advancing side material. This material did not rotate around the pin and the lower level of deformation resulted in a larger grain size.

The primary advantage of this explanation is that it provides a plausible explanation for the production of the onion-ring structure.

The marker technique for friction stir welding provides data on the initial and final positions of the marker in the welded material. The flow of material is then reconstructed from these positions. Detailed material flow field during friction stir welding can also be calculated from theoretical considerations based on fundamental scientific principles. Material flow calculations are routinely used in numerous engineering applications. Calculation of material flow fields in friction stir welding can be undertaken both using comprehensive numerical simulations or simple but insightful analytical equations. The comprehensive models for the calculation of material flow fields also provide important information such as geometry of the stir zone and the torque on the tool. The numerical simulations have shown the ability to correctly predict the results from marker experiments and the stir zone geometry observed in friction stir welding experiments.

Generation and Flow Of heat

For any welding process it is, in general, desirable to increase the travel speed and minimise the heat input as this will increase productivity and possibly reduce the impact of welding on the mechanical properties of the weld. At the same time it is necessary to ensure that the temperature around the tool is sufficiently high to permit adequate material flow and prevent flaws or tool damage.

When the traverse speed is increased, for a given heat input, there is less time for heat to conduct ahead of the tool and the thermal gradients are larger. At some point the speed will be so high that the material ahead of the tool will be too cold, and the flow stress too high, to permit adequate material movement, resulting in flaws or tool fracture. If the "hot zone" is too large then there is scope to increase the traverse speed and hence productivity.

The welding cycle can be split into several stages during which the heat flow and thermal profile will be different:

- *Dwell*. The material is preheated by a stationary, rotating tool to achieve a sufficient temperature ahead of the tool to allow the traverse. This period may also include the plunge of the tool into the workpiece.

- *Transient heating*. When the tool begins to move there will be a transient period where the heat production and temperature around the tool will alter in a complex manner until an essentially steady-state is reached.

- *Pseudo steady-state.* Although fluctuations in heat generation will occur the thermal field around the tool remains effectively constant, at least on the macroscopic scale.

- *Post steady-state.* Near the end of the weld heat may "reflect" from the end of the plate leading to additional heating around the tool.

Heat generation during friction-stir welding arises from two main sources: friction at the surface of the tool and the deformation of the material around the tool. The heat generation is often assumed to occur predominantly under the shoulder, due to its greater surface area, and to be equal to the power required to overcome the contact forces between the tool and the workpiece. The contact condition under the shoulder can be described by sliding friction, using a friction coefficient μ and interfacial pressure P, or sticking friction, based on the interfacial shear strength at an appropriate temperature and strain rate. Mathematical approximations for the total heat generated by the tool shoulder Q_{total} have been developed using both sliding and sticking.

where ω is the angular velocity of the tool, $R_{shoulder}$ is the radius of the tool shoulder and R_{pin} that of the pin. Several other equations have been proposed to account for factors such as the pin but the general approach remains the same.

A major difficulty in applying these equations is determining suitable values for the friction co-efficient or the interfacial shear stress. The conditions under the tool are both extreme and very difficult to measure. To date, these parameters have been used as "fitting parameters" where the model works back from measured thermal data to obtain a reasonable simulated thermal field. While this approach is useful for creating process models to predict, for example, residual stresses it is less useful for providing insights into the process itself.

Applications

The FSW process is currently patented by TWI in most industrialised countries and licensed for over 183 users. Friction stir welding and its variants friction stir spot welding and friction stir processing are used for the following industrial applications: shipbuilding and offshore, aerospace, automotive, rolling stock for railways, general fabrication, robotics, and computers.

Shipbuilding and Offshore

Two Scandinavian aluminium extrusion companies were the first to apply FSW commercially to the manufacture of fish freezer panels at Sapa in 1996, as well as deck panels and helicopter landing platforms at Marine Aluminium Aanensen. Marine Aluminium Aanensen subsequently merged with Hydro Aluminium Maritime to become Hydro Marine Aluminium. Some of these freezer panels are now produced by Riftec and Bayards. In 1997 two-dimensional friction stir welds in the hydrodynamically flared bow section of the hull of the ocean viewer vessel *The Boss* were produced at Research Foundation Institute with the first portable FSW machine. The *Super Liner Ogasawara* at Mitsui Engineering and Shipbuilding is the largest friction stir welded ship so far. The *Sea Fighter* of Nichols Bros and the *Freedom* class Littoral Combat Ships contain prefabricated panels by the FSW fabricators Advanced Technology and Friction Stir Link, Inc. respectively. The *Houbei* class missile boat has friction stir welded rocket launch containers of China Friction Stir Centre. HMNZS *Rotoiti* in New Zealand has

FSW panels made by Donovans in a converted milling machine. Various companies apply FSW to armor plating for amphibious assault ships

Friction stir welding was used to prefabricate the aluminium panels of the Super Liner Ogasawara at Mitsui Engineering and Shipbuilding

Aerospace

Longitudinal and circumferential friction stir welds are used for the Falcon 9 rocket booster tank at the SpaceX factory

United Launch Alliance applies FSW to the Delta II, Delta IV, and Atlas V expendable launch vehicles, and the first of these with a friction stir welded Interstage module was launched in 1999. The process is also used for the Space Shuttle external tank, for Ares I and for the Orion Crew Vehicle test article at NASA as well as Falcon 1 and Falcon 9 rockets at SpaceX. The toe nails for ramp of Boeing C-17 Globemaster III cargo aircraft by Advanced Joining Technologies and the cargo barrier beams for the Boeing 747 Large Cargo Freighter were the first commercially produced aircraft parts. FAA approved wings and fuselage panels of the Eclipse 500 aircraft were made at Eclipse Aviation, and this company delivered 259 friction stir welded business jets, before they were forced into Chapter 7 liquidation. Floor panels for Airbus A400M military aircraft are now made by Pfalz Flugzeugwerke and Embraer used FSW for the Legacy 450 and 500 Jets Friction stir welding also is employed for fuselage panels on the Airbus A380. BRÖTJE-Automation GmbH uses friction stir welding – through the DeltaN FS® system – for gantry production machines developed for the aerospace sector as well as other industrial applications.

Automotive

The centre tunnel of the Ford GT is made from two aluminium extrusions friction stir welded to a bent aluminium sheet and houses the fuel tank

Aluminium engine cradles and suspension struts for stretched Lincoln Town Car were the first automotive parts that were friction stir at Tower Automotive, who use the process also for the engine tunnel of the Ford GT. A spin-off of this company is called Friction Stir Link, Inc. and successfully exploits the FSW process, e.g. for the flatbed trailer "Revolution" of Fontaine Trailers. In Japan FSW is applied to suspension struts at Showa Denko and for joining of aluminium sheets to galvanized steel brackets for the boot (trunk) lid of the Mazda MX-5. Friction stir spot welding is successfully used for the bonnet (hood) and rear doors of the Mazda RX-8 and the boot lid of the Toyota Prius. Wheels are friction stir welded at Simmons Wheels, UT Alloy Works and Fundo. Rear seats for the Volvo V70 are friction stir welded at Sapa, HVAC pistons at Halla Climate Control and exhaust gas recirculation coolers at Pierburg. Tailor welded blanks are friction stir welded for the Audi R8 at Riftec. The B-column of the Audi R8 Spider is friction stir welded from two extrusions at Hammerer Aluminium Industries in Austria.

Railways

The high-strength low-distortion body of Hitachi's A-train *British Rail Class 395* is friction stir welded from longitudinal aluminium extrusions

Since 1997 roof panels were made from aluminium extrusions at Hydro Marine Aluminium with a bespoke 25m long FSW machine, e.g. for DSB class SA-SD trains of Alstom LHB Curved side and roof panels for the Victoria line trains of London Underground, side panels for Bombardier's Electrostar trains at Sapa Group and side panels for Alstom's British Rail Class 390 Pendolino trains are made at Sapa Group Japanese commuter and express A-trains, and British Rail Class 395 trains are friction stir welded by Hitachi, while Kawasaki applies friction stir spot welding to roof panels and Sumitomo Light Metal produces Shinkansen floor panels. Innovative FSW floor panels are made by Hammerer Aluminium Industries in Austria for the Stadler KISS double decker rail cars, to obtain an internal height of 2 m on both floors and for the new car bodies of the Wuppertal Suspension Railway.

Heat sinks for cooling high-power electronics of locomotives are made at Sykatek, EBG, Austerlitz Electronics, EuroComposite, Sapa and Rapid Technic, and are the most common application of FSW due to the excellent heat transfer.

Fabrication

The lids of 50-mm-thick copper canisters for nuclear waste are attached to the cylinder by friction stir welding at SKB

Friction stir processed knives by MegaStir

Façade panels and cathode sheets are friction stir welded at AMAG and Hammerer Aluminium Industries including friction stir lap welds of copper to aluminium. Bizerba meat slicers, Ökolüfter HVAC units and Siemens X-ray vacuum vessels are friction stir welded at Riftec. Vacuum valves and vessels are made by FSW at Japanese and Swiss companies. FSW is also used for the encapsulation of nuclear waste at SKB in 50-mm-thick copper canisters. Pressure vessels from ø1m semispherical forgings of 38.1mm thick aluminium alloy 2219 at Advanced Joining Technologies and Lawrence Livermore Nat Lab. Friction stir processing is applied to ship propellers at Friction Stir Link, Inc. and to hunting knives by DiamondBlade. Bosch uses it in Worcester for the production of heat exchangers.

Robotics

KUKA Robot Group has adapted its KR500-3MT heavy-duty robot for friction stir welding via the DeltaN FS tool. The system made its first public appearance at the EuroBLECH show in November 2012.

Personal Computers

Apple applied friction stir welding on the 2012 iMac to effectively join the bottom to the back of the device.

Electron Beam Welding

Electron beam welding (EBW) is a fusion welding process in which a beam of high-velocity electrons is applied to two materials to be joined. The workpieces melt and flow together as the kinetic energy of the electrons is transformed into heat upon impact. EBW is often performed under vacuum conditions to prevent dissipation of the electron beam.

History

Electron beam welding was developed by the German physicist Karl-Heinz Steigerwald, who was at the time working on various electron beam applications. Steigerwald conceived and developed the first practical electron beam welding machine, which began operation in 1958. American inventor James T. Russell has also been credited with designing and building the first electron-beam welder.

Electron beam welder

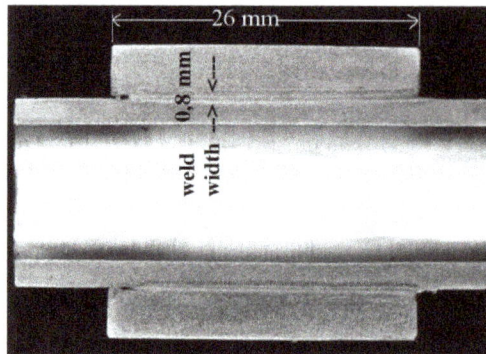

Deep narrow weld

Physics of Electron Beam Heating

Electrons are elementary particles possessing a mass $m = 9.1 \cdot 10^{-31}$ kg and a negative electrical charge $e = 1.6 \cdot 10^{-19}$ C. They exist either bound to an atomic nucleus, as conduction electrons in the atomic lattice of metals, or as free electrons in vacuum.

Free electrons in vacuum can be accelerated, with their orbits controlled by electric and magnetic fields. In this way narrow beams of electrons carrying high kinetic energy can be formed, which upon collision with atoms in solids transform their kinetic energy into heat. Electron beam welding provides excellent welding conditions because it involves:

- Strong electric fields, which can accelerate electrons to a very high speed. Thus, the electron beam can carry high power, equal to the product of beam current and accelerating voltage. By increasing the beam current and the accelerating voltage, the beam power can be increased to practically any desired value.

- Using magnetic lenses, by which the beam can be shaped into a narrow cone and focused to a very small diameter. This allows for a very high surface power density on the surface to be welded. Values of power density in the crossover (focus) of the beam can be as high as $10^4 - 10^6$ W/mm².

- Shallow penetration depths in the order of hundredths of a millimeter. This allows for a very high volumetric power density, which can reach values of the order $10^5 - 10^7$ W/mm³. Consequently, the temperature in this volume increases extremely rapidly, $10^8 - 10^{10}$ K/s.

The effectiveness of the electron beam depends on many factors. The most important are the physical properties of the materials to be welded, especially the ease with which they can be melted or vaporize under low-pressure conditions. Electron beam welding can be so intense that loss of material due to evaporation or boiling during the process must be taken into account when welding. At lower values of surface power density (in the range of about 10^3 W/mm²) the loss of material by evaporation is negligible for most metals, which is favorable for welding. At higher power density, the material affected by the beam can totally evaporate in a very short time; this is no longer electron beam welding; it is electron beam machining.

Beam Formation

Cathode - the source of free electrons

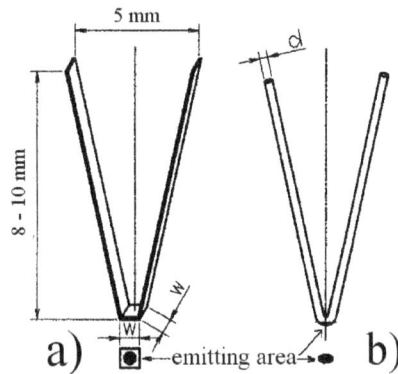

Tungsten cathodes: strap - wire

Conduction electrons (those not bound to the nucleus of atoms) move in a crystal lattice of metals with velocities distributed according to Gauss's law and depending on temperature. They cannot leave the metal unless their kinetic energy (in eV) is higher than the potential barrier at the metal surface. The number of electrons fulfilling this condition increases exponentially with increasing temperature of the metal, following Richardson's rule.

As a source of electrons for electron beam welders, the material must fulfill certain requirements:

- to achieve high power density in the beam, the emission current density [A/mm²], hence the working temperature, should be as high as possible,

- to keep evaporation in vacuum low, the material must have a low enough vapour pressure at the working temperature.

- The emitter must be mechanically stable, not chemically sensitive to gases present in the vacuum atmosphere (like oxygen and water vapour), easily available, etc.

These and other conditions limit the choice of material for the emitter to metals with high melting points, practically to only two: tantalum and tungsten. With tungsten cathodes, emission current densities about 100 mA/mm² can be achieved, but only a small portion of the emitted electrons takes part in beam formation, depending on the electric field produced by the anode and control electrode voltages. The type of cathode most frequently used in electron beam welders is made of a tungsten strip, about 0.05 mm thick, shaped as shown in Fig. 1a. The appropriate width of the strip depends on the highest required value of emission current. For the lower range of beam power, up to about 2 kW, the width w=0.5 mm is appropriate.

Acceleration of electrons, current control

Electrons emitted from the cathode possess very low energy, only a few eV. To give them the required high speed, they are accelerated by a strong electric field applied between the emitter and another, positively charged, electrode, namely the anode. The accelerating field must also navigate the electrons to form a narrow converging "bundle" around the axis. This can be achieved by an electric field in the proximity of the emitting cathode surface which has, a radial addition as well as an axial component, forcing the electrons in the direction of the axis. Due to this effect, the electron beam converges to some minimum diameter in a plane close to the anode.

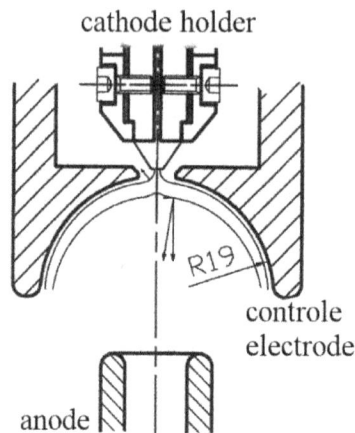

Beam generator

For practical applications the power of the electron beam must, of course, be controllable. This can be accomplished by another electric field produced by another cathode negatively charged with respect to the first.

At least this part of electron gun must be evacuated to "high" vacuum, to prevent "burning" the cathode and the emergence of electrical discharges.

After leaving the anode, the divergent electron beam does not have a power density sufficient for welding metals and has to be focused. This can be accomplished by a magnetic field produced by electric current in a cylindrical coil.

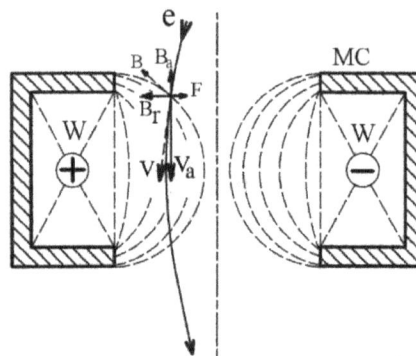

Magnetic lens

The focusing effect of a rotationally symmetrical magnetic field on the trajectory of electrons is the result of the complicated influence of a magnetic field on a moving electron. This effect is a force proportional to the induction B of the field and electron velocity v. The vector product of the radial component of induction B_r and axial component of velocity v_a is a force perpendicular to those vectors, causing the electron to move around the axis. Additional effect of this motion in the same magnetic field is another force F oriented radially to the axis, which is responsible for the focusing effect of the magnetic lens. The resulting trajectory of electrons in the magnetic lens is a curve similar to a helix. In this context it should be mentioned that variations of focal length (exciting current) cause a slight rotation of the beam cross-section.

Beam deflection system

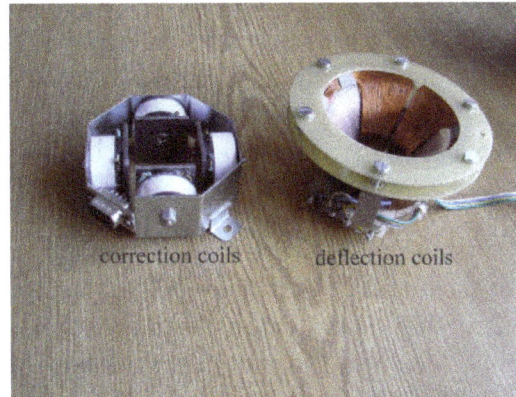

Correction & deflection coils

As mentioned above, the beam spot should be very precisely positioned with respect to the joint to be welded. This is commonly accomplished mechanically by moving the workpiece with respect to the electron gun, but sometimes it is preferable to deflect the beam instead. Most often a system of four coils positioned symmetrically around the gun axis behind the focusing lens, producing a magnetic field perpendicular to the gun axis, is used for this purpose.

There are more practical reasons why the most appropriate deflection system is used in TV CRT or PC monitors. This applies to both the deflecting coils as well as to the necessary electronics. Such a system enables not only "static" deflection of the beam for the positioning purposes mentioned above, but also precise and fast dynamic control of the beam spot position by a computer. This makes it possible, e.g.:

- to weld joints of complicated geometry,

- to create image-enlarged pictures of objects in the working chamber on TV or PC monitors.

Both possibilities find many useful applications in electron beam welding practice.

Penetration of Electron Beam During Welding

To explain the capability of the electron beam to produce deep and narrow welds, the process of "penetration" must be explained. First of all, the process for a "single" electron can be considered.

When electrons from the beam impact the surface of a solid, some of them may be reflected (as "backscattered" electrons), while others penetrate the surface, where they collide with the particles of the solid. In non-elastic collisions they lose their kinetic energy. It has been proved, both theoretically and experimentally, that they can "travel" only a very small distance below the surface before they transfer all their kinetic energy into heat. This distance is proportional to their initial energy and inversely proportional to the density of the solid. Under conditions usual in welding practice the "travel distance" is on the order of hundredths of a millimeter. Just this fact enables, under certain conditions, fast beam penetration.

The heat contribution of single electrons is very small, but the electrons can be accelerated by very high voltages, and by increasing their number (the beam current) the power of the beam can be increased to any desired value. By focusing the beam onto a small diameter on the surface of a solid object, values of planar power density as high as 10^4 up to 10^7 W/mm^2 can be reached. Because electrons transfer their energy into heat in a very thin layer of the solid, as explained above, the power density in this volume can be extremely high. The volume density of power in the small volume in which the kinetic energy of the electrons is transformed into heat can reach values of the order $10^5 - 10^7$ W/mm^3. Consequently, the temperature in this volume increases extremely rapidly, by $10^8 - 10^9$ K/s.

The effect of the electron beams under such circumstances depends on several conditions, first of all on the physical properties of the material. Any material can be melted, or even evaporated, in a very short time. Depending on conditions, the intensity of evaporation may vary, from negligible to essential. At lower values of surface power density (in the range of about 10^3 W/mm^2) the loss of material by evaporation is negligible for most metals, which is favorable for welding. At higher power density, the material affected by the beam can totally evaporate in a very short time; this no longer electron beam welding; it is electron beam machining.

Results of the Electron Beam Application

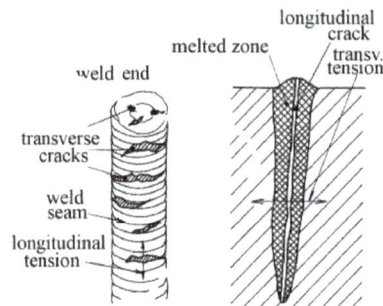

Various forms of melted zone

The results of the beam application depend on several factors: Many experiments and innumerable practical applications of electron beam in welding technology prove that the effect of the beam, i.e. the size and shape of the zone influenced by the beam depends on:

(1) Beam power – The power of the beam [W] is the product of the accelerating voltage [kV] and beam current [mA], parameters easily measurable and precisely controllable. The power is controlled by the beam current at constant accelerating voltage, usually the highest accessible.

(2) Power density (focusing of the beam) – The power density at the spot of incidence of the beam with the workpiece depends on factors like the size of the electron source on the cathode, the optical quality of the accelerating electric lens and the focusing magnetic lens, alignment of the beam, the value of the accelerating voltage, and the focal length. All these factors (except the focal length) depend on the design of the machine.

(3) Welding speed – The construction of the welding equipment should enable adjustment of the relative speed of motion of the workpiece with respect to the beam in wide enough limits, e.g., between 2 and 50 mm/s.

(4) Material properties, and in some cases also on

(5) Geometry (shape and dimensions) of the joint.

The final effect of the beam depends on the particular combination of these parameters.

- Action of the beam at low power density or over a very short time results in melting only a thin surface layer.

- A defocused beam does not penetrate, and the material at low welding speeds is heated only by conduction of the heat from the surface, producing a hemispherical melted zone.

- At high power density and low speed, a deeper and slightly conical melted zone is produced.

- In the case of very high power density, the beam (well focused) penetrates deeper, in proportion to its total power.

Welded membranes

The Welding Process

Weldability

For welding thin-walled parts, appropriate welding aids are generally needed. Their construction must provide perfect contact of the parts and prevent their movement during welding. Usually they have to be designed individually for a given workpiece.

Not all materials can be welded by an electron beam in a vacuum. This technology cannot be applied to materials with high vapour pressure at the melting temperature, like zinc, cadmium, magnesium and practically all non-metals.

Another limitation to weldability may be the change of material properties induced by the welding process, such as a high speed of cooling. As detailed discussion of this matter exceeds the scope of this article, the reader is recommended to seek more information in the appropriate literature.

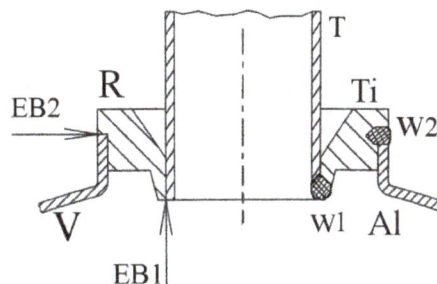
Titanium-to-aluminium joints

Joining Dissimilar Materials

It is often not possible to join two metal components by welding, i.e. to melt part of both in the vicinity of the joint, if the two materials have very different properties from their alloy, due to the creation of brittle, inter-metallic compounds. This situation cannot be changed, even by electron beam heating in vacuum, but this nevertheless makes it possible to realize joints meeting high demands for mechanical compactness and that are perfectly vacuum-tight. The principal approach is not to melt both parts, but only the one with the lower melting point, while the other remains solid. The advantage of electron beam welding is its ability to localize heating to a precise point and to control exactly the energy needed for the process. A high-vacuum atmosphere substantially contributes to a positive result. A general rule for construction of joints to be made this way is that the part with the lower melting point should be directly accessible for the beam.

Possible Problems and Limitations

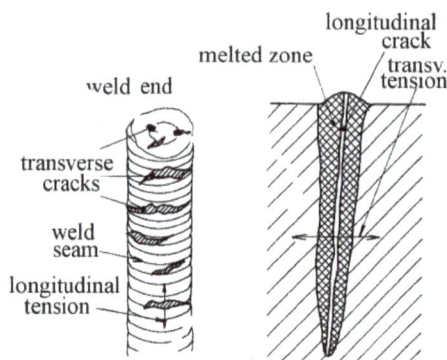

Cracks in weld

The material melted by the beam shrinks during cooling after solidification, which may have unwanted consequences like cracking, deformation and changes of shape, depending on conditions.

The butt weld of two plates results in bending of the weldment because more material has been melted at the head than at the root of the weld. This effect is of course not as substantial as in arc welding.

Another potential danger is the emergence of cracks in the weld. If both parts are rigid, the shrinkage of the weld produces high stress in the weld which may lead to cracks if the material is brittle (even if only after remelting by welding). The consequences of weld contraction should always be considered when constructing the parts to be welded.

Electron Beam Welding Equipment

Since the publication of the first practical electron beam welding equipment by Steigerwald in 1958, electron beam welding has spread rapidly in all branches of engineering where welding can be applied. To cover the various requirements, countless welder types have been designed, differing in construction, working space volume, workpiece manipulators and beam power. Electron beam generators (electron guns) designed for welding applications can supply beams with power ranging from a few watts up to about one hundred kilowatts. "Micro-welds" of tiny components can be realized, as well as deep welds up to 300 mm (or even more if needed). Vacuum working chambers of various design may have a volume of only a few liters, but vacuum chambers with the volume of several hundreds cubic meters have also been built.

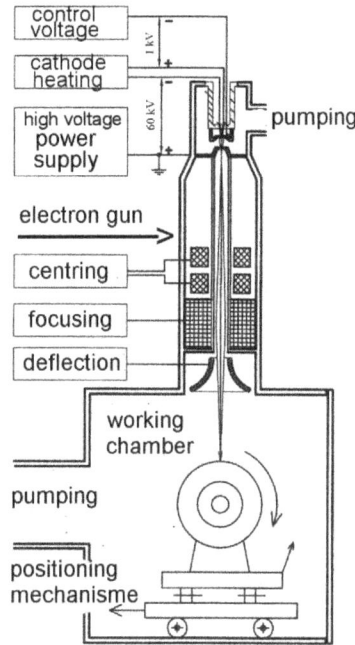

Electron beam welder

Specifically, the equipment comprises:

1 Electron gun, generating the electron beam,

2 Working chamber, mostly evacuated to "low" or "high" vacuum,

3 Workpiece manipulator (positioning mechanism),

4 Power supply and control and monitoring electronics.

In the electron gun, the free electrons are gained by thermo-emission from a hot metal strap (or wire). They are then accelerated and formed into a narrow convergent beam by an electric field produced by three electrodes: the electron emitting strap, the cathode connected to the negative pole of the high (accelerating) voltage power supply (30 - 200 kV) and the positive high voltage electrode, the anode. There is a third electrode charged negatively with respect to the cathode, called the Wehnelt or control electrode. Its negative potential controls the portion of emitted electrons entering into the accelerating field, i.e., the electron beam current.

After passing the anode opening, the electrons move with constant speed in a slightly divergent cone. For technological applications the divergent beam has to be focused, which is realized by the magnetic field of a coil, the magnetic focusing lens.

For proper functioning of the electron gun, it is necessary that the beam be perfectly adjusted with respect to the optical axes of the accelerating electrical lens and the magnetic focusing lens. This can be done by applying a magnetic field of some specific radial direction and strength perpendicular to the optical axis before the focusing lens. This is usually realized by a simple correction system consisting of two pairs of coils. By adjusting the currents in these coils any required correcting field can be produced.

After passing the focusing lens, the beam can be applied for welding, either directly or after being deflected by a deflection system. This consists of two pairs of coils, one for each X and Y direction. These can be used for "static" or "dynamic" deflection. Static deflection is useful for exact positioning of the beam by welding. Dynamic deflection is realized by supplying the deflection coils with currents which can be controlled by the computer. This opens new possibilities for electron beam applications, like surface hardening or annealing, exact beam positioning, etc.

The fast deflection system can also be applied (if provided with appropriate electronics) for imaging and engraving. In this case the equipment is operated like a scanning electron microscope, with a resolution of about 0,1 mm (limited by the beam diameter). In a similar mode the fine computer-controlled beam can "write" or "draw" a picture on the metal surface by melting a thin surface layer.

Since the appearance of the first electron beam welding machines at the end of the 1950s, the application of electron beam welding spread rapidly into industry and research in all highly developed countries. Up to now, uncountable numbers of various types of electron beam equipment have been designed and realized. In most of them the welding takes place in a working vacuum chamber in a high or low vacuum environment.

The vacuum working chamber may have any desired volume, from a few liters up to hundreds of cubic meters. They can be provided with electron guns supplying an electron beam with any required power up to 100 kW, or even more if needed. In micro-electron beam devices, components with dimensions in tenths of a millimeter can be precisely welded. In welders with electron beams of high enough power, welds up to 300 mm deep can be realized.

There are also welding machines in which the electron beam is brought out of vacuum into the atmosphere. With such equipment very large objects can be welded without huge working chambers.

Electron beam welding can never be "hand-manipulated", even if not realized in vacuum, as there is always strong X-radiation. The relative motion of the beam and the workpiece is most often achieved by rotation or linear travel of the workpiece. In some cases the welding is realized by moving the beam with the help of a computer-controlled deflection system. Workpiece manipulators are mostly designed individually to meet the specific requirements of the welding equipment.

Electron beam equipment must be provided with an appropriate power supply for the beam generator. The accelerating voltage may be chosen between 30 and 200 kV. Usually it is about 60 or 150 kV, depending on various conditions. With rising voltage the technical problems and the price of the equipment rapidly increase, hence, whenever it is possible a lower voltage of about 60 kV is to be chosen. The maximum power of the high voltage supply depends on the maximum depth of weld required.

The high-voltage equipment must also supply the low voltage, above 5 V, for the cathode heating, and negative voltage up to about 1000 V for the control electrode.

The electron gun also needs low-voltage supplies for the correction system, the focusing lens, and the deflection system. The last mentioned may be very complex if it is to provide computer-controlled imaging, engraving, or similar beam applications.

Laser Beam Welding

A robot performs remote fibre laser welding.

Laser beam welding (LBW) is a welding technique used to join multiple pieces of metal through the use of a laser. The beam provides a concentrated heat source, allowing for narrow, deep welds and high welding rates. The process is frequently used in high volume applications using automation, such as in the automotive industry. It is based on keyhole or penetration mode welding.

Operation

Like electron beam welding (EBW), laser beam welding has high power density (on the order of $1 MW/cm^2$) resulting in small heat-affected zones and high heating and cooling rates. The spot size of the laser can vary between 0.2 mm and 13 mm, though only smaller sizes are used for welding. The depth of penetration is proportional to the amount of power supplied, but is also dependent on the location of the focal point: penetration is maximized when the focal point is slightly below the surface of the workpiece

A continuous or pulsed laser beam may be used depending upon the application. Millisecond-long pulses are used to weld thin materials such as razor blades while continuous laser systems are employed for deep welds.

LBW is a versatile process, capable of welding carbon steels, HSLA steels, stainless steel, aluminum, and titanium. Due to high cooling rates, cracking is a concern when welding high-carbon steels. The weld quality is high, similar to that of electron beam welding. The speed of welding is proportional to the amount of power supplied but also depends on the type and thickness of the workpieces. The high power capability of gas lasers make them especially suitable for high volume applications. LBW is particularly dominant in the automotive industry.

Some of the advantages of LBW in comparison to EBW are as follows:

- the laser beam can be transmitted through air rather than requiring a vacuum

- the process is easily automated with robotic machinery

- x-rays are not generated

- LBW results in higher quality welds

A derivative of LBW, laser-hybrid welding, combines the laser of LBW with an arc welding method such as gas metal arc welding. This combination allows for greater positioning flexibility, since GMAW supplies molten metal to fill the joint, and due to the use of a laser, increases the welding speed over what is normally possible with GMAW. Weld quality tends to be higher as well, since the potential for undercutting is reduced.

Equipment

Automation and CAM

Although laser beam welding can be accomplished by hand, most systems are automated use a system of computer aided manufacturing based on computer aided designs. Laser welding can also be coupled with milling to form a finished part.

Recently the RepRap project, which historically worked on fused filament fabrication, expanded to development of open source laser welding systems. Such systems have been fully characterized and can be used in a wide scale of applications while reducing conventional manufacturing costs.

Lasers

- The two types of lasers commonly used are solid-state lasers (especially ruby lasers and Nd:YAG lasers) and gas lasers.

- The first type uses one of several solid media, including synthetic ruby (chromium in aluminum oxide), neodymium in glass (Nd:glass), and the most common type, neodymium in yttrium aluminum garnet (Nd:YAG).

- Gas lasers use mixtures of gases such as helium, nitrogen, and carbon dioxide (CO_2 laser) as a medium.

- Regardless of type, however, when the medium is excited, it emits photons and forms the laser beam.

Solid State Laser

Solid-state lasers operate at wavelengths on the order of 1 micrometer, much shorter than gas lasers, and as a result require that operators wear special eyewear or use special screens to prevent retina damage. Nd:YAG lasers can operate in both pulsed and continuous mode, but the other types are limited to pulsed mode. The original and still popular solid-state design is a single crystal shaped as a rod approximately 20 mm in diameter and 200 mm long, and the ends are ground flat.

This rod is surrounded by a flash tube containing xenon or krypton. When flashed, a pulse of light lasting about two milliseconds is emitted by the laser. Disk shaped crystals are growing in popularity in the industry, and flashlamps are giving way to diodes due to their high efficiency. Typical power output for ruby lasers is 10–20 W, while the Nd:YAG laser outputs between 0.04–6,000 W. To deliver the laser beam to the weld area, fiber optics are usually employed.

Gas Laser

Gas lasers use high-voltage, low-current power sources to supply the energy needed to excite the gas mixture used as a lasing medium. These lasers can operate in both continuous and pulsed mode, and the wavelength of the CO_2 gas laser beam is 10.6 μm, deep infrared, i.e. 'heat'. Fiber optic cable absorbs and is destroyed by this wavelength, so a rigid lens and mirror delivery system is used. Power outputs for gas lasers can be much higher than solid-state lasers, reaching 25 kW.

Fiber Laser

In fiber lasers, the gain medium is the optical fiber itself. They are capable of power up to 50 kW and are increasingly being used for robotic industrial welding.

Laser Beam Delivery

Modern laser beam welding machines can be grouped into two types. In the traditional type, the laser output is moved to follow the seam. This is usually achieved with a robot. In many modern applications, remote laser beam welding is used. In this method, the laser beam is moved along the seam with the help of a laser scanner, so that the robotic arm does not need to follow the seam any more. The advantages of remote laser welding are the higher speed and the higher precision of the welding process.

Robot Welding

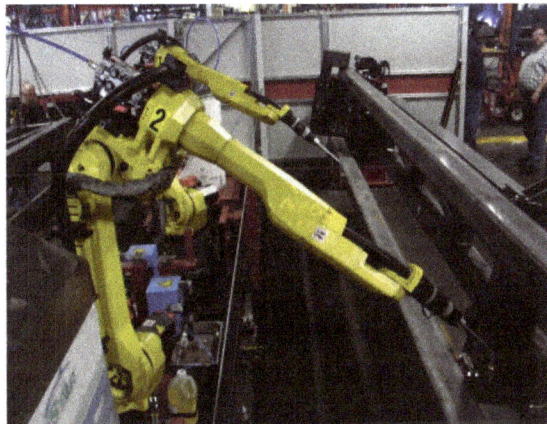

A set of six-axis robots used for welding.

Robot welding is the use of mechanized programmable tools (robots), which completely automate a welding process by both performing the weld and handling the part. Processes such as gas metal arc welding, while often automated, are not necessarily equivalent to robot welding, since a human operator sometimes prepares the materials to be welded. Robot welding is commonly used for resistance spot welding and arc welding in high production applications, such as the automotive industry.

Robot welding is a relatively new application of robotics, even though robots were first introduced into US industry during the 1960s. The use of robots in welding did not take off until the 1980s, when the automotive industry began using robots extensively for spot welding. Since then, both the number of robots used in industry and the number of their applications has grown greatly. In 2005, more than 120,000 robots were in use in North American industry, about half of them for welding. Growth is primarily limited by high equipment costs, and the resulting restriction to high-production applications. In 2014, FANUC America Corp. introduced a low cost arc welding robot to provide small manufacturers with a cost-effective robotic arc welding solution.

Robot arc welding has begun growing quickly just recently, and already it commands about 20% of industrial robot applications. The major components of arc welding robots are the manipulator or the mechanical unit and the controller, which acts as the robot's "brain". The manipulator is what makes the robot move, and the design of these systems can be categorized into several common types, such as SCARA and cartesian coordinate robot, which use different coordinate systems to direct the arms of the machine.

The robot may weld a pre-programmed position, be guided by machine vision, or by a combination of the two methods. However, the many benefits of robotic welding have proven to make it a technology that helps many original equipment manufacturers increase accuracy, repeat-ability, and throughput

The technology of signature image processing has been developed since the late 1990s for analyzing electrical data in real time collected from automated, robotic welding, thus enabling the optimization of welds.

Stud Welding

Slab base weld nuts

Stud welding is a technique similar to flash welding where a fastener or specially formed nut is welded onto another metal part, typically a base metal or substrate. The fastener can take different forms, but typically fall under threaded, unthreaded or tapped. The bolts may be automatically fed into the spot welder. Weld nuts generally have a flange with small nubs that melt to form the weld. *Weld studs* are used in stud welding systems.

Capacitor discharge weld studs range from 14 gauge to 3/8" diameter. They can come in many different lengths ranging from 1/4" to 5" and larger. The tip on the weld end of the stud serves a twofold purpose:

1. It acts as a timing device to keep the stud off the base material

2. It disintegrates when the trigger is pulled on the gun.

When the tip disintegrates, it melts and helps solidify the weld to the base material.

Arc studs range from a #8 to $1\frac{1}{4}$" diameter. The lengths are variable from 3/8" to 60" (for deformed bars). Arc studs are typically loaded with an aluminum flux ball on the weld end which aids in the welding process.

Stud welding, also known as "drawn arc stud welding", joins a stud and another piece of metal together by heating both parts with an arc. The stud is usually joined to a flat plate by using the stud as one of the electrodes. The polarity used in stud welding depends on the type of metal being used. Welding aluminum, for example, would usually require direct-current electrode positive (DCEP). Welding steel would require direct-current electrode negative (DCEN).

Stud welding uses a type of flux called a ferrule, a ceramic ring which concentrates the heat generated, prevents oxidation and retains the molten metal in the weld zone. The ferrule is broken off of the fastener after the weld is completed. This lack of marring on the side opposite the fastener is what differentiates stud welding from other fastening processes.

Portable stud welding machines are available. Welders can also be automated, with controls for arcing and applying pressure. Stud welding is very versatile. Typical applications include automobile bodies, electrical panels, shipbuilding and building construction. Shipbuilding is one of the oldest uses of stud welding. Stud welding revolutionized the shipbuilding industry. All other manufacturing industries can also use stud welding for a variety of purposes.

A type of stud welding called capacitor-discharge (CD) stud welding differs from regular stud welding in that capacitor-discharge welding does not require flux. The weld time is shorter, enabling the weld to bond with little oxidation and no need for heat concentration. It also allows for small-diameter studs to be welded to thin,lightweight materials. This process uses a direct-current arc from a capacitor. The weld time in this process is between 1 and 6 milliseconds. Capacitor discharge stud welding with the latest equipment can create a weld without burn through showing on the opposite side of very thin metals. CD stud welding is often used for smaller diameter studs and pins, as well as on non-standard materials and for accuracy. On the other hand, arc stud welding is primarily for structural purposes and larger diameter weld studs.

Standards

Among the standards quoted in the list of welding codes, the following apply:

- ISO 13918 - Welding - Studs and ceramic ferrules for arc stud welding

- ISO 14555 - Welding - Arc stud welding of metallic materials

Upset Welding

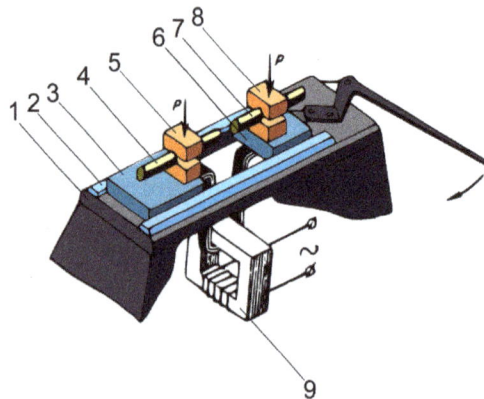

upset(butt)welding machine

Upset welding (UW)/resistance butt welding is a welding technique that produces coalescence simultaneously over the entire area of abutting surfaces or progressively along a joint, by the heat obtained from resistance to electric current through the area where those surfaces are in contact. . Pressure is applied before heating is started and is maintained throughout the heating period. The equipment used for upset welding is very similar to that used for flash welding. It can be used only if the parts to be welded are equal in cross-sectional area. The abutting surfaces must be very carefully prepared to provide for proper heating. The difference from flash welding is that the parts are clamped in the welding machine and force is applied bringing them tightly together. High-amperage current is then passed through the joint, which heats the abutting surfaces. When they have been heated to a suitable forging temperature an upsetting force is applied and the current is stopped. The high temperature of the work at the abutting surfaces plus the high pressure causes coalescence to take place. After cooling, the force is released and the weld is completed.

Diffusion Bonding

Diffusion bonding is a solid-state welding technique used in metalworking, capable of joining similar and dissimilar metals. It operates on the principle of solid-state diffusion, wherein the atoms of two solid, metallic surfaces intersperse themselves over time. This is typically accomplished at an elevated temperature, approximately one-half of the absolute melting temperature of the materials. Diffusion bonding is usually implemented by applying high pressure, in conjunction with necessarily high temperature, to the materials to be welded; the technique is most commonly used to weld "sandwiches" of alternating layers of thin metal foil, and metal wires or filaments. Currently, the diffusion bonding method is widely used in the joining of high-strength and refractory metals within the aerospace and nuclear industries.

History

The act of diffusion welding is centuries old. This can be found in the form of "filled gold," a technique used to bond gold and copper for use in jewelry and other applications. In order to create filled gold, smiths would begin by hammering out an amount of solid gold into a thin sheet of gold foil. This film was then placed on top of a copper substrate and weighted down. Finally, using a process known as "hot-pressure welding" or HPW, the weight/copper/gold-film assembly was placed inside an oven and heated until the gold film was sufficiently bonded to the copper substrate.

Characteristics

Diffusion bonding involves no liquid fusion, and often no filler metal. No weight is added to the total, and the join tends to exhibit both the strength and temperature resistance of the base metal(s). The materials endure no, or very little, plastic deformation. Very little residual stress is introduced, and there is no contamination from the bonding process. It may be performed on a join surface of theoretically any size with no increase in processing time; practically speaking, the surface tends to be limited by the pressure required and physical limitations. It may be performed with similar and dissimilar metals, reactive and refractory metals, or pieces of varying thicknesses.

Diffusion bonding is most often used for jobs either difficult or impossible to weld by other means, due to its relatively high cost. Examples include welding materials normally impossible to join via liquid fusion, such as zirconium and beryllium; materials with very high melting points such as tungsten; alternating layers of different metals which must retain strength at high temperatures; and very thin, honeycombed metal foil structures.

Temperature Dependence

Steady state diffusion is determined by the amount of diffusion flux that passes through the cross sectional area of the mating surfaces. Fick's first law of diffusion states:

$$J = -D(dC / dx)$$

where J is the diffusion flux, D is a diffusion coefficient, and dC/dx is the concentration gradient through the materials in question. The negative sign is a product of the gradient. Another form of Fick's law states:

$$J = M / (At)$$

where M is defined as either the mass or amount of atoms being diffused, A is the cross sectional area, and t is the time required. Equating the two equations and rearranging, we achieve the following result:

$$t = -(1 / D)(M / A)(dC / dx)^{-1}$$

As mass and area are constant for a given joint, time required is largely dependent on the concentration gradient, which changes by only incremental amounts through the joint, and the diffusion coefficient. The diffusion coefficient is determined by the equation:

$$t = -(1/D)(M/A)(dC/dx)^{-1}$$

where Q_d is the activation energy for diffusion, R is the universal gas constant, T is the temperature experienced during the process in Kelvin, and D_o is a temperature-independent preexponential that depends on the materials being joined. For a given joint, the only term in this equation within control is temperature.

Processes

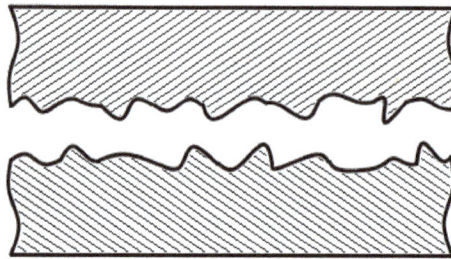

Animation of Diffusion Bonding Process

When joining to materials of similar crystalline structure, diffusion bonding is performed by clamping the two pieces to be welded with their surfaces abutting each other. Prior to welding, these surfaces must be machined to as smooth a finish as economically viable, and kept as free from chemical contaminants or other detritus as possible. Any intervening material between the two metallic surfaces may prevent adequate diffusion of material. Once clamped, pressure and heat are applied to the components, usually for many hours. The surfaces are heated either in a furnace, or via electrical resistance. Pressure can be applied using a hydraulic press at temperature; this method allows for exact measurements of load on the parts. In cases where the parts must have no temperature gradient, differential thermal expansion can be used to apply load. By fixturing parts using a low-expansion metal (i.e. Molybdenum) the parts will supply their own load by expanding more than the fixture metal at temperature. Alternative methods for applying pressure include the use of dead weights, differential gas pressure between the two surfaces, and high-pressure autoclaves. Diffusion bonding must be done in a vacuum or inert gas environment when using metals that have strong oxide layers (i.e. copper). Surface treatment including polishing, etching, and cleaning as well as diffusion pressure and temperature are important factors regarding to process of diffusion bounding.

At the microscopic level, diffusion bonding occurs in three simplified stages:

- Before the surfaces completely contact, asperities (very small surface defects) on the two surfaces contact and plastically deform. As these asperities deform, they interlink, forming interfaces between the two surfaces.

- Elevated temperature and pressure causes accelerated creep in the materials; grain boundaries and raw material migrate and gaps between the two surfaces are reduced to isolated pores.

- Material begins to diffuse across the boundary of the abutting surfaces, blending this material boundary and creating a bond.

Benefits

- The bounded surface have the same physical and mechanical properties as the base material. Once we have finished the jointing, we could also perform the test of the jointing materials, for example, tensile testing.

- The diffusion bounding process is able to produce a high quality joints in which case no discontinuity and porosity exists in the interface. In other words, we are able to sand, manufacturing and heat the material.

- The diffusion bounding is able to help us to build high precision components with complex shapes. Also, diffusion is flexible.

- The diffusion bounding method can be used wildly, joining either similar or dissimilar materials, and also important in processing composite materials.

- The process is not extremely hard to approach and the cost to perform the diffusion bounding is not high.

- The material under diffusion is able to reduce the plastic deformation.

Applicability

Animation of sheet forming process using diffusion welding

Diffusion bonding is primarily used to create intricate forms for the electronics, aerospace, and nuclear industries. Since this form of bonding takes a considerable amount of time compared to other joining techniques such as explosion welding, parts are made in small quantities, and often fabrication is mostly automated. However, due to different requirements, some of the time interval could be accomplished in few minutes. In an attempt to reduce fastener count, labor costs, and part count, diffusion bonding, in conjunction with superplastic forming, is also used when creating complex sheet metal forms. Multiple sheets are stacked atop one another and bonded in specific sections. The stack is then placed into a mold and gas pressure expands the sheets to fill the mold. This is often done using titanium or aluminum alloys for parts needed in the aerospace industry.

References

- Houldcroft, P. T. (1973) [1967]. "Chapter 3: Flux-Shielded Arc Welding". Welding Processes. Cambridge University Press. p. 23. ISBN 0-521-05341-2.

- Cary, Howard B.; Helzer, Scott C. (2005), Modern Welding Technology, Upper Saddle River, New Jersey: Pearson Education, ISBN 0-13-113029-3

- American Welding Society (2004). Welding handbook, welding processes Part 1. Miami Florida: American Welding Society. ISBN 0-87171-729-8.

- Cary, Howard B.; Helzer, Scott C. (2005). Modern welding technology. Upper Saddle River, New Jersey: Pearson Education. ISBN 0-13-113029-3.

- Minnick, William H. (1996). Gas tungsten arc welding handbook. Tinley Park, Illinois: Goodheart–Willcox Company. ISBN 1-56637-206-2.

- Pires, J Roberto; Loureiro, Altino; Bolmsjö, Gunnar (2005). Welding Robots: Technology, System Issues and Application. New York: Springer. p. 11. ISBN 1-85233-953-5.

- Larry F. Jeffus (2002). Welding: Principles and Applications. Cengage Learning. p. 694. ISBN 9781401810467. Retrieved April 18, 2014.

- S. R. Deb; S. Deb (2010). Robotics Technology and Flexible Automation. Tata McGraw-Hill Education. p. 491. ISBN 9780070077911. Retrieved April 18, 2014.

- George F. Schrader; Ahmad K. Elshennawy (2000). Manufacturing Processes and Materials. SME. p. 311. ISBN 9780872635173. Retrieved April 18, 2014.

- D. Lohwasser and Z. Chen: "Friction stir welding — From basics to applications" Woodhead Publishing 2010, Chapter 5, Pages 118–163, ISBN 978-1-84569-450-0.

- Schultz, Helmut (1993). Electron beam welding. Cambridge, England: Woodhead Publishing/The Welding Institute. ISBN 1-85573-050-2.

- Cary, Howard B. and Scott C. Helzer (2005). Modern Welding Technology. Upper Saddle River, New Jersey: Pearson Education. Page 316. ISBN 0-13-113029-3.

- Schrader, George F.; Elshennway, Ahmad K. Manufacturing Processes and MAterials (4th illustrated ed.). pp. 319–320. ISBN 0872635171.

- Chawla, Krishan K. Composite Materials: Science and Engineering. Materials research and engineering (2nd illustrated ed.). p. 174. ISBN 0387984097.

- Callister, William D. Jr.; Rethwisch, David G. (2014). Materials Science and Engineering: An Introduction, 9th ed. John Wiley and Sons Inc. pp. 143–151. ISBN 978-1-118-32457-8.

Processes of Welding

The process of binding metal to metal by using electricity is known as arc welding. The other processes of welding are electric resistance welding, filet weld, plasma cutting, fusion welding, explosion welding, cold welding etc. The aspects elucidated in this chapter are of vital importance, and provide a better understanding of welding.

Arc Welding

Arc welding is a process that is used to join metal to metal by using electricity to create heat enough to melt metal, and the melted metals when cool result in a binding of the metals. It is a type of welding that uses a welding power supply to create an electric arc between an electrode and the base material to melt the metals at the welding point. They can use either direct (DC) or alternating (AC) current, and consumable or non-consumable electrodes. The welding region is usually protected by some type of shielding gas, vapor, or slag. Arc welding processes may be manual, semi-automatic, or fully automated. First developed in the late part of the 19th century, arc welding became commercially important in shipbuilding during the Second World War. Today it remains an important process for the fabrication of steel structures and vehicles.

Power Supplies

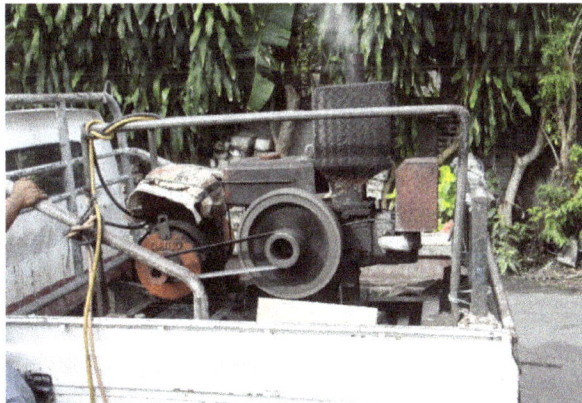

A diesel powered welding generator (the electric generator is on the left) as used in Indonesia.

To supply the electrical energy necessary for arc welding processes, a number of different power supplies can be used. The most common classification is constant current power supplies and constant voltage power supplies. In arc welding, the voltage is directly related to the length of the arc, and the current is related to the amount of heat input. Constant current power supplies are most often used for manual welding processes such as gas tungsten arc welding and shielded metal arc welding, because they maintain a relatively constant current even as the voltage varies. This is important because in manual welding, it can be difficult to hold the electrode perfectly steady, and as a result, the arc length and thus voltage tend to fluctuate. Constant voltage power supplies hold the voltage constant and vary the current, and as a result, are most often used for automated welding processes such as gas metal arc welding, flux cored arc welding, and submerged arc welding. In these processes, arc length is kept constant, since any fluctuation in the distance between the wire and the base material is quickly rectified by a large change in current. For example, if the wire and the base material get too close, the current will rapidly increase, which in turn causes the heat to increase and the tip of the wire to melt, returning it to its original separation distance.

The direction of current used in arc welding also plays an important role in welding. Consumable electrode processes such as shielded metal arc welding and gas metal arc welding generally use direct current, but the electrode can be charged either positively or negatively. In welding, the positively charged anode will have a greater heat concentration (around 60%) and, as a result, changing the polarity of the electrode affects weld properties. If the electrode is positively charged, it will melt more quickly, increasing weld penetration and welding speed. Alternatively, a negatively charged electrode results in more shallow welds. Non-consumable electrode processes, such as gas tungsten arc welding, can use either type of direct current (DC), as well as alternating current (AC). With direct current however, because the electrode only creates the arc and does not provide filler material, a positively charged electrode causes shallow welds, while a negatively charged electrode makes deeper welds. Alternating current rapidly moves between these two, resulting in medium-penetration welds. One disadvantage of AC, the fact that the arc must be re-ignited after every zero crossing, has been addressed with the invention of special power units that produce a square wave pattern instead of the normal sine wave, eliminating low-voltage time after the zero crossings and minimizing the effects of the problem.

Duty cycle is a welding equipment specification which defines the number of minutes, within a 10-minute period, during which a given arc welder can safely be used. For example, an 80 A welder with a 60% duty cycle must be "rested" for at least 4 minutes after 6 minutes of continuous welding. Failure to observe duty cycle limitations could damage the welder. Commercial- or professional-grade welders typically have a 100% duty cycle.

Consumable Electrode Methods

One of the most common types of arc welding is shielded metal arc welding (SMAW), which is also known as manual metal arc welding (MMAW) or stick welding. An electric current is used to strike an arc between the base material and a consumable electrode rod or *stick*. The electrode rod is made of a material that is compatible with the base material being welded and is covered with a flux that gives off vapors that serve as a shielding gas and provide a layer of slag, both of which protect the weld area from atmospheric contamination. The electrode core itself acts as filler material, making a separate filler unnecessary. The process is very versatile, requiring little operator training and inexpensive equipment. However, weld times are rather slow, since the consumable electrodes must be frequently replaced and because slag, the residue from the flux, must be chipped away after welding. Furthermore, the process is generally limited to welding ferrous materials, though specialty electrodes have made possible the welding of cast iron, nickel, aluminium, copper and other metals. The versatility of the method makes it popular in a number of applications including repair work and construction.

Shielded metal arc welding

Gas metal arc welding (GMAW), commonly called *MIG* (for *metal/inert-gas*), is a semi-automatic or automatic welding process with a continuously fed consumable wire acting as both electrode and filler metal, along with an inert or semi-inert shielding gas flowed around the wire to protect the weld site from contamination. Constant voltage, direct current power source is most commonly used with GMAW, but constant current alternating current are used as well. With continuously fed filler electrodes, GMAW offers relatively high welding speeds, however the more complicated equipment reduces convenience and versatility in comparison to the SMAW process. Originally developed for welding aluminium and other non-ferrous materials in the 1940s, GMAW was soon economically applied to steels. Today, GMAW is commonly used in industries such as the automobile industry for its quality, versatility and speed. Because of the need to maintain a stable shroud of shielding gas around the weld site, it can be problematic to use the GMAW process in areas of high air movement such as outdoors.

Flux-cored arc welding (FCAW) is a variation of the GMAW technique. FCAW wire is actually a fine metal tube filled with powdered flux materials. An externally supplied shielding gas is sometimes used, but often the flux itself is relied upon to generate the necessary protection from the atmosphere. The process is widely used in construction because of its high welding speed and portability.

Submerged arc welding (SAW) is a high-productivity welding process in which the arc is struck beneath a covering layer of granular flux. This increases arc quality, since contaminants in the atmosphere are blocked by the flux. The slag that forms on the weld generally comes off by itself and, combined with the use of a continuous wire feed, the weld deposition rate is high. Working conditions are much improved over other arc welding processes since the flux hides the arc and no smoke is produced. The process is commonly used in industry, especially for large products. As the arc is not visible, it is typically automated. SAW is only possible in the 1F (flat fillet), 2F (horizontal fillet), and 1G (flat groove) positions.

Non-consumable Electrode Methods

Gas tungsten arc welding (GTAW), or *tungsten/inert-gas* (TIG) welding, is a manual welding process that uses a non-consumable electrode made of tungsten, an inert or semi-inert gas mixture, and a separate filler material. Especially useful for welding thin materials, this method is characterized by a stable arc and high quality welds, but it requires significant operator skill and can only be accomplished at relatively low speeds. It can be used on nearly all weldable metals, though it is most often applied to stainless steel and light metals. It is often used when quality welds are extremely important, such as in bicycle, aircraft and naval applications.

A related process, plasma arc welding, also uses a tungsten electrode but uses plasma gas to make the arc. The arc is more concentrated than the GTAW arc, making transverse control more critical and thus generally restricting the technique to a mechanized process. Because of its stable current, the method can be used on a wider range of material thicknesses than can the GTAW process and is much faster. It can be applied to all of the same materials as GTAW except magnesium; automated welding of stainless steel is one important application of the process. A variation of the process is plasma cutting, an efficient steel cutting process.

Other arc welding processes include atomic hydrogen welding, carbon arc welding, electroslag welding, electrogas welding, and stud arc welding.

Corrosion Issues

Some materials, notably high-strength steels, aluminium, and titanium alloys, are susceptible to hydrogen embrittlement. If the electrodes used for welding contain traces of moisture, the water decomposes in the heat of the arc and the liberated hydrogen enters the lattice of the material, causing its brittleness. Stick electrodes for such materials, with special low-hydrogen coating, are delivered in sealed moisture-proof packaging. New electrodes can be used straight from the can, but when moisture absorption may be suspected, they have to be dried by baking (usually at 450 to 550 °C or 840 to 1,020 °F) in a drying oven. Flux used has to be kept dry as well.

Some austenitic stainless steels and nickel-based alloys are prone to intergranular corrosion. When subjected to temperatures around 700 °C (1,300 °F) for too long a time, chromium reacts with carbon in the material, forming chromium carbide and depleting the crystal edges of chromium, impairing their corrosion resistance in a process called sensitization. Such sensitized steel undergoes corrosion in the areas near the welds where the temperature-time was favorable for forming the carbide. This kind of corrosion is often termed weld decay.

Knifeline attack (KLA) is another kind of corrosion affecting welds, impacting steels stabilized by niobium. Niobium and niobium carbide dissolves in steel at very high temperatures. At some cooling regimes, niobium carbide does not precipitate, and the steel then behaves like unstabilized steel, forming chromium carbide instead. This affects only a thin zone several millimeters wide in the very vicinity of the weld, making it difficult to spot and increasing the corrosion speed. Structures made of such steels have to be heated in a whole to about 1,000 °C (1,830 °F), when the chromium carbide dissolves and niobium carbide forms. The cooling rate after this treatment is not important.

Filler metal (electrode material) improperly chosen for the environmental conditions can make them corrosion-sensitive as well. There are also issues of galvanic corrosion if the electrode composition is sufficiently dissimilar to the materials welded, or the materials are dissimilar themselves. Even between different grades of nickel-based stainless steels, corrosion of welded joints can be severe, despite that they rarely undergo galvanic corrosion when mechanically joined.

Safety Issues

Welding safety checklist

Welding can be a dangerous and unhealthy practice without the proper precautions; however, with the use of new technology and proper protection the risks of injury or death associated with welding can be greatly reduced.

Heat, Fire, and Explosion Hazard

Because many common welding procedures involve an open electric arc or flame, the risk of burns from heat and sparks is significant. To prevent them, welders wear protective clothing in the form of heavy leather gloves and protective long sleeve jackets to avoid exposure to extreme heat, flames, and sparks. The use of compressed gases and flames in many welding processes also pose an explosion and fire risk; some common precautions include limiting the amount of oxygen in the air and keeping combustible materials away from the workplace.

Eye Damage

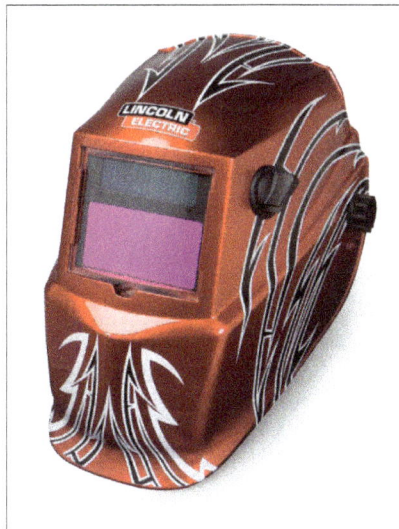

Auto darkening welding hood with 90×110 mm cartridge and 3.78×1.85 in viewing area

Exposure to the brightness of the weld area leads to a condition called arc eye in which ultraviolet light causes inflammation of the cornea and can burn the retinas of the eyes. Welding goggles and helmets with dark face plates—much darker than those in sunglasses or oxy-fuel goggles—are worn to prevent this exposure. In recent years, new helmet models have been produced featuring a face plate which automatically self-darkens electronically. To protect bystanders, transparent welding curtains often surround the welding area. These curtains, made of a polyvinyl chloride plastic film, shield nearby workers from exposure to the UV light from the electric arc.

Inhaled Matter

Welders are also often exposed to dangerous gases and particulate matter. Processes like flux-cored arc welding and shielded metal arc welding produce smoke containing particles of various types of oxides. The size of the particles in question tends to influence the toxicity of the fumes, with smaller particles presenting a greater danger. Additionally, many processes produce various gases (most commonly carbon dioxide and ozone, but others as well) that can prove dangerous if ventilation is inadequate.

Interference With Pacemakers

Certain welding machines which use a high frequency alternating current component have been found to affect pacemaker operation when within 2 meters of the power unit and 1 meter of the weld site.

History

Nikolay Benardos

While examples of forge welding go back to the Bronze Age and the Iron Age, arc welding did not come into practice until much later.

In 1800 Sir Humphry Davy discovered the short pulsed electric arcs. Independently a Russian physicist Vasily Petrov discovered the continuous electric arc in 1802 and subsequently proposed its possible practical applications, including welding. Arc welding was first developed when Nikolai Benardos presented arc welding of metals using a carbon electrode at the International Exposition of Electricity, Paris in 1881, which was patented together with Stanisław Olszewski in 1887. In the same year, French electrical inventor Auguste de Méritens invented also a carbon arc welding method, patented in 1881, which was successfully used for welding lead in the manufacture of lead-acid batteries. The advances in arc welding continued with the invention of metal electrodes in the late 19th century by a Russian, Nikolai Slavyanov (1888), and an American, C. L. Coffin. Around 1900, A. P. Strohmenger released in Britain a coated metal electrode which gave a more stable arc. In 1905 Russian scientist Vladimir Mitkevich proposed the usage of three-phase electric arc for welding. In 1919, alternating current welding was invented by C.J. Holslag but did not become popular for another decade.

Competing welding processes such as resistance welding and oxyfuel welding were developed during this time as well; but both, especially the latter, faced stiff competition from arc welding especially after metal coverings (known as flux) for the electrode, to stabilize the arc and shield the base material from impurities, continued to be developed.

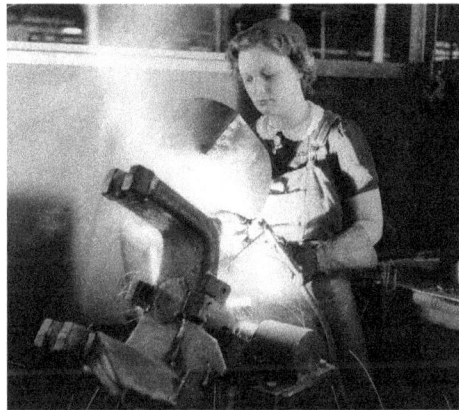

A young woman arc welding in a munitions factory in Australia in 1943.

During World War I welding started to be used in shipbuilding in Great Britain in place of riveted steel plates. The Americans also became more accepting of the new technology when the process allowed them to repair their ships quickly after a German attack in the New York Harbor at the beginning of the war. Arc welding was first applied to aircraft during the war as well, and some German airplane fuselages were constructed using this process. In 1919, the British shipbuilder Cammell Laird started construction of merchant ship, the "Fullagar", with an entirely welded hull; she was launched in 1921.

During the 1920s, major advances were made in welding technology, including the 1920 introduction of automatic welding in which electrode wire was continuously fed. Shielding gas became a subject receiving much attention as scientists attempted to protect welds from the effects of oxygen and nitrogen in the atmosphere. Porosity and brittleness were the primary problems and the solutions that developed included the use of hydrogen, argon, and helium as welding atmospheres. During the following decade, further advances allowed for the welding of reactive metals such as aluminum and magnesium. This, in conjunction with developments in automatic welding, alternating current, and fluxes fed a major expansion of arc welding during the 1930s and then during World War II.

During the middle of the century, many new welding methods were invented. Submerged arc welding was invented in 1930 and continues to be popular today. In 1932 a Russian, Konstantin Khrenov successfully implemented the first underwater electric arc welding. Gas tungsten arc welding, after decades of development, was finally perfected in 1941 and gas metal arc welding followed in 1948, allowing for fast welding of non-ferrous materials but requiring expensive shielding gases. Using a consumable electrode and a carbon dioxide atmosphere as a shielding gas, it quickly became the most popular metal arc welding process. In 1957, the flux-cored arc welding process debuted in which the self-shielded wire electrode could be used with automatic equipment, resulting in greatly increased welding speeds. In that same year, plasma arc welding was invented. Electroslag welding was released in 1958 and was followed by its cousin, electrogas welding, in 1961.

Oxy-fuel Welding and Cutting

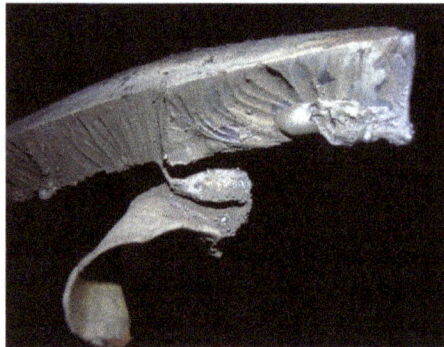

Side of metal, cut by oxygen - propane cutting torch

A cutting torch is used to cut a steel pipe.

Oxy-fuel welding (commonly called oxyacetylene welding, oxy welding, or gas welding in the U.S.) and oxy-fuel cutting are processes that use fuel gases and oxygen to weld and cut metals, respectively. French engineers Edmond Fouché and Charles Picard became the first to develop oxygen-acetylene welding in 1903. Pure oxygen, instead of air, is used to increase the flame temperature to allow localized melting of the workpiece material (e.g. steel) in a room environment. A common propane/air flame burns at about 2,250 K (1,980 °C; 3,590 °F), a propane/oxygen flame burns at about 2,526 K (2,253 °C; 4,087 °F), an oxyhydrogen flame burns at 2,800 °C (5,070 °F), and an acetylene/oxygen flame burns at about 3,773 K (3,500 °C; 6,332 °F).

Oxy-fuel is one of the oldest welding processes, besides forge welding. In recent decades it has been obsolesced in most all industrial uses due to various arc welding methods offering more consistent mechanical weld properties and faster application. Gas welding is still used for metal-based artwork and in smaller home based shops, as well as situations where accessing electricity (e.g., via an extension cord or portable generator) would present difficulties.

In oxy-fuel welding, a welding torch is used to weld metals. Welding metal results when two pieces are heated to a temperature that produces a shared pool of molten metal. The molten pool is generally supplied with additional metal called filler. Filler material depends upon the metals to be welded.

In oxy-fuel cutting, a torch is used to heat metal to its kindling temperature. A stream of oxygen is then trained on the metal, burning it into a metal oxide that flows out of the kerf as slag.

Torches that do not mix fuel with oxygen (combining, instead, atmospheric air) are not considered oxy-fuel torches and can typically be identified by a single tank (oxy-fuel cutting requires two isolated supplies, fuel and oxygen). Most metals cannot be melted with a single-tank torch. Consequently, single-tank torches are typically suitable for soldering and brazing but not for welding.

Uses

Oxy-gas torches are or have been used for:

- Welding metal

- Cutting metal

- Depositing metal to build up a surface, as in hardfacing

- Also, oxy-hydrogen flames are used:

  o in stone working for "flaming" where the stone is heated and a top layer crackles and breaks. A steel circular brush is attached to an angle grinder and used to remove the first layer leaving behind a bumpy surface similar to hammered bronze.

  o in the glass industry for "fire polishing".

  o in jewelry production for "water welding" using a water torch (an oxyhydrogen torch whose gas supply is generated immediately by electrolysis of water).

  o in automotive repair, removing a seized bolt.

  o formerly, to heat lumps of quicklime to obtain a bright white light called limelight, in theatres or optical ("magic") lanterns.

- formerly, in platinum works, as platinum is fusible only in the oxyhydrogen flame- and in an electric furnace.

In short, oxy-fuel equipment is quite versatile, not only because it is preferred for some sorts of iron or steel welding but also because it lends itself to brazing, braze-welding, metal heating (for annealing or tempering, bending or forming), rust or scale removal, the loosening of corroded nuts and bolts, and is a ubiquitous means of cutting ferrous metals.

Apparatus

The apparatus used in gas welding consists basically of an oxygen source and a fuel gas source (usually contained in cylinders), two pressure regulators and two flexible hoses (one for each cylinder), and a torch. This sort of torch can also be used for soldering and brazing. The cylinders are often carried in a special wheeled trolley.

There have been examples of oxyhydrogen cutting sets with small (scuba-sized) gas cylinders worn on the user's back in a backpack harness, for rescue work and similar.

There are also examples of pressurized liquid fuel cutting torches, usually using gasoline. These are used for their increased portability.

Regulator

The regulator ensures that pressure of the gas from the tanks matches the required pressure in the hose. The flow rate is then adjusted by the operator using needle valves on the torch. Accurate flow control with a needle valve relies on a constant inlet pressure.

Most regulators have two stages. The first stage is a fixed-pressure regulator, which releases gas from the cylinder at a constant intermediate pressure, despite the pressure in the cylinder falling as the gas in it is consumed. This is similar to the first stage of a scuba-diving regulator. The adjustable second stage of the regulator controls the pressure reduction from the intermediate pressure to the low outlet pressure. The regulator has two pressure gauges, one indicating cylinder pressure, the other indicating hose pressure. The adjustment knob of the regulator is sometimes roughly calibrated for pressure, but an accurate setting requires observation of the gauge.

Some simpler or cheaper oxygen-fuel regulators have only a single stage regulator, or only a single gauge. A single-stage regulator will tend to allow a reduction in outlet pressure as the cylinder is emptied, requiring manual readjustment. For low-volume users, this is an acceptable simplification. Welding regulators, unlike simpler LPG heating regulators, retain their outlet (hose) pressure gauge and do not rely on the calibration of the adjustment knob. The cheaper single-stage regulators may sometimes omit the cylinder contents gauge, or replace the accurate dial gauge with a cheaper and less precise "rising button" gauge.

Gas Hoses

The hoses are designed for use in welding and cutting metal. A double-hose or twinned design can be used, meaning that the oxygen and fuel hoses are joined together. If separate hoses are used, they should be clipped together at intervals approximately 3 feet (1 m) apart, although that is not recommended for cutting applications, because beads of molten metal given off by the process can become lodged between the hoses where they are held together, and burn through, releasing the pressurised gas inside, which in the case of fuel gas usually ignites.

The hoses are color-coded for visual identification. The color of the hoses varies between countries. In the United States, the oxygen hose is green, and the fuel hose is red. In the UK and other countries, the oxygen hose is blue (black hoses may still be found on old equipment), and the acetylene (fuel) hose is red. If liquefied petroleum gas (LPG) fuel, such as propane, is used, the fuel hose should be orange, indicating that it is compatible with LPG. LPG will damage an incompatible hose, including most acetylene hoses.

The threaded connectors on the hoses are handed to avoid accidental mis-connection: the thread on the oxygen hose is right-handed (as normal), while the fuel gas hose has a left-handed thread. The left-handed threads also have an identifying groove cut into their nuts.

Gas-tight connections between the flexible hoses and rigid fittings are made by using crimped hose clips or ferrules, often referred to as 'O' clips, over barbed spigots. The use of worm-drive hose clips or Jubilee clips is specifically forbidden in the UK and other countries.

Non-return Valve

Acetylene is not just flammable, in certain conditions it is explosive. Although it has an upper flammability limit in air of 81%, acetylene's explosive decomposition behaviour makes this irrelevant. If a detonation wave enters the acetylene tank, the tank will be blown apart by the decomposition. Ordinary check valves that normally prevent back flow cannot stop a detonation wave because they are not capable of closing before the wave passes around the gate. For that reason a flashback arrestor is needed. It is designed to operate before the detonation wave makes it from the hose side to the supply side.

Between the regulator and hose, and ideally between hose and torch on both oxygen and fuel lines, a flashback arrestor and/or non-return valve (check valve) should be installed to prevent flame or oxygen-fuel mixture being pushed back into either cylinder and damaging the equipment or causing a cylinder to explode.

European practice is to fit flashback arrestors at the regulator and check valves at the torch. US practice is to fit both at the regulator.

The flashback arrestor prevents shock waves from downstream coming back up the hoses and entering the cylinder, possibly rupturing it, as there are quantities of fuel/oxygen mixtures inside parts of the equipment (specifically within the mixer and blowpipe/nozzle) that may explode if the equipment is incorrectly shut down, and acetylene decomposes at excessive pressures or temperatures. In case the pressure wave has created a leak downstream of the flashback arrestor, it will remain switched off until someone resets it.

Check Valve

A check valve lets gas flow in one direction only. It is usually a chamber containing a ball that is pressed against one end by a spring. Gas flow one way pushes the ball out of the way, and a lack of flow or a reverse flow allows the spring push the ball into the inlet, blocking it. A check valve is not designed to block a shock wave. The shock wave could occur while the ball is so far from the inlet that the wave will get past the ball before it can reach its off position.

Torch

The torch is the tool that the welder holds and manipulates to make the weld. It has a connection and valve for the fuel gas and a connection and valve for the oxygen, a handle for the welder to grasp, and a mixing chamber (set at an angle) where the fuel gas and oxygen mix, with a tip where the flame forms. Two basic types of torches are positive pressure type and low pressure or injector type.

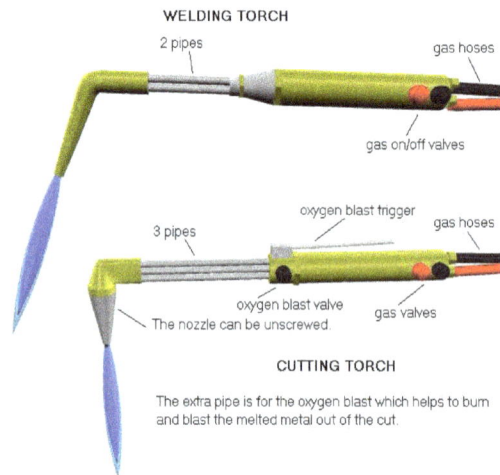

WELDING TORCH
2 pipes
gas hoses
gas on/off valves

oxygen blast trigger
gas hoses
3 pipes
oxygen blast valve
gas valves
The nozzle can be unscrewed.

CUTTING TORCH
The extra pipe is for the oxygen blast which helps to burn and blast the melted metal out of the cut.

The top torch is a welding torch and the bottom is a cutting torch

Welding Torch

A welding torch head is used to weld metals. It can be identified by having only one or two pipes running to the nozzle, no oxygen-blast trigger, and two valve knobs at the bottom of the handle letting the operator adjust the oxygen and fuel flow respectively.

Cutting Torch

A cutting torch head is used to cut materials. It is similar to a welding torch, but can be identified by the oxygen blast trigger or lever.

When cutting, the metal is first heated by the flame until it is cherry red. Once this temperature is attained, oxygen is supplied to the heated parts by pressing the oxygen-blast trigger. This oxygen reacts with the metal, forming iron oxide and producing heat. It is the heat that continues the cutting process. The cutting torch only heats the metal to start the process; further heat is provided by the burning metal.

The melting point of the iron oxide is around half that of the metal being cut. As the metal burns, it immediately turns to liquid iron oxide and flows away from the cutting zone. However, some of the iron oxide remains on the workpiece, forming a hard "slag" which can be removed by gentle tapping and/or grinding.

Rose Bud Torch

A rose bud torch is used to heat metals for bending, straightening, etc. where a large area needs to be heated. It is so-called because the flame at the end looks like a rose bud. A welding torch can also be used to heat small areas such as rusted nuts and bolts.

Injector Torch

A typical oxy-fuel torch, called an equal-pressure torch, merely mixes the two gases. In an injector torch, high-pressure oxygen comes out of a small nozzle inside the torch head which drags the fuel gas along with it, using the venturi effect.

Fuels

Oxy-fuel processes may use a variety of fuel gases, the most common being acetylene. Other gases that may be used are propylene, liquified petroleum gas (LPG), propane, natural gas, hydrogen, and MAPP gas. Many brands use different kinds of gases in their mixes.

Acetylene

Acetylene generator as used in Bali by a reaction of calcium carbide with water. This is used where acetylene cylinders are not available. The term 'Las Karbit' means acetylene (carbide) welding in Indonesian.

Acetylene is the primary fuel for oxy-fuel welding and is the fuel of choice for repair work and general cutting and welding. Acetylene gas is shipped in special cylinders designed to keep the gas dissolved. The cylinders are packed with porous materials (e.g. kapok fibre, diatomaceous earth, or (formerly) asbestos), then filled to around 50% capacity with acetone, as acetylene is acetone soluble. This method is necessary because above 207 kPa (30 lbf/in²) (absolute pressure) acetylene is unstable and may explode.

There is about 1700 kPa (250 psi) pressure in the tank when full. Acetylene when combined with oxygen burns at 3200 °C to 3500 °C (5800 °F to 6300 °F), highest among commonly used gaseous fuels. As a fuel acetylene's primary disadvantage, in comparison to other fuels, is high cost.

As acetylene is unstable at a pressure roughly equivalent to 33 feet/10 meters underwater, water submerged cutting and welding is reserved for hydrogen rather than acetylene.

Compressed gas cylinders containing oxygen and MAPP gas.

Gasoline

Oxy-gasoline, also known as oxy-petrol, torches have been found to perform very well, especially where bottled gas fuel is not available or difficult to transport to the worksite. Tests showed that an oxy-gasoline torch can cut steel plate up to 0.5 in (13 mm) thick at the same rate as oxy-acetylene. In plate thicknesses greater than 0.5 in (13 mm) the cutting rate was better than oxy-acetylene; at 4.5 in (110 mm) it was three times faster.

The gasoline is fed either from a pressurised tank (whose pressure can be hand-pumped or fed from a gas cylinder). OR from a non pressurised tank with the fuel being drawn into the torch by venturi action by the pressurised oxygen flow. Another low cost approach commonly used by jewelry makers in Asia is using air bubbled through a gasoline container by a foot-operated air pump, and burning the fuel-air mixture in a specialized welding torch.

Hydrogen

Hydrogen has a clean flame and is good for use on aluminium. It can be used at a higher pressure than acetylene and is therefore useful for underwater welding and cutting. It is a good type of flame to use when heating large amounts of material. The flame temperature is high, about 2,000 °C for hydrogen gas in air at atmospheric pressure, and up to 2800 °C when pre-mixed in a 2:1 ratio with pure oxygen (oxyhydrogen). Hydrogen is not used for welding steels and other ferrous materials, because it causes hydrogen embrittlement.

For some oxyhydrogen torches the oxygen and hydrogen are produced by electrolysis of water in an apparatus which is connected directly to the torch. Types of this sort of torch:

- The oxygen and the hydrogen are led off the electrolysis cell separately and are fed into the two gas connections of an ordinary oxy-gas torch. This happens in the water torch, which is sometimes used in small torches used in making jewelry and electronics.

- The mixed oxygen and hydrogen are drawn from the electrolysis cell and are led into a special torch designed to prevent flashback.

MPS and MAPP Gas

Methylacetylene-propadiene (MAPP) *gas* and *LPG gas* are similar fuels, because LPG gas is liquefied petroleum gas mixed with MPS. It has the storage and shipping characteristics of LPG and has a heat value a little less than acetylene. Because it can be shipped in small containers for sale at retail stores, it is used by hobbyists and large industrial companies and shipyards because it does not polymerize at high pressures — above 15 psi or so (as acetylene does) and is therefore much less dangerous than acetylene. Further, more of it can be stored in a single place at one time, as the increased compressibility allows for more gas to be put into a tank. MAPP gas can be used at much higher pressures than acetylene, sometimes up to 40 or 50 psi in high-volume oxy-fuel cutting torches which can cut up to 12-inch-thick (300 mm) steel. Other welding gases that develop comparable temperatures need special procedures for safe shipping and handling. MPS and MAPP are recommended for cutting applications in particular, rather than welding applications.

On 31 April 2008 the Petromont Varennes plant closed its methylacetylene/propadiene crackers. As they were the only North American plant making MAPP gas, many substitutes were introduced by the companies who had repackaged the Dow and Varennes product(s) - most of these substitutes are propylene.

Propylene and Fuel Gas

Propylene is used in production welding and cutting. It cuts similarly to propane. When propylene is used, the torch rarely needs tip cleaning. There is often a substantial advantage to cutting with an injector torch rather than an equal-pressure torch when using propylene. Quite a few North American suppliers have begun selling propylene under proprietary trademarks such as FG2 and Fuel-Max.

Butane, Propane and Butane/Propane Mixes

Butane, like propane, is a saturated hydrocarbon. Butane and propane do not react with each other and are regularly mixed. Butane boils at 0.6 °C. Propane is more volatile, with a boiling point of -42 °C. Vaporization is rapid at temperatures above the boiling points. The calorific (heat) values of both are almost equal. Both are thus mixed to attain the vapor pressure that is required by the end user and depending on the ambient conditions. If the ambient temperature is very low, propane is preferred to achieve higher vapor pressure at the given temperature.

Propane does not burn as hot as acetylene in its inner cone, and so it is rarely used for welding. Propane, however, has a very high number of BTUs per cubic foot in its outer cone, and so with the right torch (injector style) can make a faster and cleaner cut than acetylene, and is much more useful for heating and bending than acetylene.

The maximum neutral flame temperature of propane in oxygen is 2,822 °C (5,112 °F).

Propane is cheaper than acetylene and easier to transport.

The Role of Oxygen

Oxygen is not the fuel. It is what chemically combines with the fuel to produce the heat for welding. This is called 'oxidation', but the more specific and more commonly used term in this context is 'combustion'. In the case of hydrogen, the product of combustion is simply water. For the other hydrocarbon fuels, water and carbon dioxide are produced. The heat is released because the molecules of the products of combustion have a lower energy state than the molecules of the fuel and oxygen. In oxy-fuel cutting, oxidation of the metal being cut (typically iron) produces nearly all of the heat required to "burn" through the workpiece.

Oxygen is usually produced elsewhere by distillation of liquified air and shipped to the welding site in high pressure vessels (commonly called "tanks" or "cylinders") at a pressure of about 21,000 kPa (3,000 lbf/in² = 200 atmospheres). It is also shipped as a liquid in Dewar type vessels (like a large Thermos jar) to places that use large amounts of oxygen.

It is also possible to separate oxygen from air by passing the air, while under pressure, through a zeolite sieve which selectively absorbs the nitrogen and lets the oxygen (and argon) pass. This gives a purity of oxygen of about 93%. This works well for brazing, however higher purity oxygen is necessary to produce a clean, slag-free kerf when cutting.

Types of Flame

The welder can adjust the oxy-acetylene flame to be carbonizing (aka reducing), neutral, or oxidizing. Adjustment is made by adding more or less oxygen to the acetylene flame. The neutral flame is the flame most generally used when welding or cutting. The welder uses the neutral flame as the starting point for all other flame adjustments because it is so easily defined. This flame is attained when welders, as they slowly open the oxygen valve on the torch body, first see only two flame zones. At that point, the acetylene is being completely burned in the welding oxygen and surrounding air. The flame is chemically neutral. The two parts of this flame are the light blue inner cone and the darker blue to colorless outer cone. The inner cone is where the acetylene and the oxygen combine. The tip of this inner cone is the hottest part of the flame. It is approximately 6,000 °F (3,300 °C) and provides enough heat to easily melt steel. In the inner cone the acetylene breaks down and partly burns to hydrogen and carbon monoxide, which in the outer cone combine with more oxygen from the surrounding air and burn.

An excess of acetylene creates a carbonizing flame. This flame is characterized by three flame zones; the hot inner cone, a white-hot "acetylene feather", and the blue-colored outer cone. This is the type of flame observed when oxygen is first added to the burning acetylene. The feather is adjusted and made ever smaller by adding increasing amounts of oxygen to the flame. A welding feather is measured as 2X or 3X, with X being the length of the inner flame cone. The unburned carbon insulates the flame and drops the temperature to approximately 5,000 °F (2,800 °C). The reducing flame is typically used for hard facing operations or backhand pipe welding techniques. The feather is caused by incomplete combustion of the acetylene to cause an excess of carbon in the flame. Some of this carbon is dissolved by the molten metal to carbonize it. The carbonizing flame will tend to remove the oxygen from iron oxides which may be present, a fact which has caused the flame to be known as a "reducing flame".

The oxidizing flame is the third possible flame adjustment. It occurs when the ratio of oxygen to acetylene required for a neutral flame has been changed to give an excess of oxygen. This flame type is observed when welders add more oxygen to the neutral flame. This flame is hotter than the other two flames because the combustible gases will not have to search so far to find the necessary amount of oxygen, nor heat up as much thermally inert carbon. It is called an oxidizing flame because of its effect on metal. This flame adjustment is generally not preferred. The oxidizing flame creates undesirable oxides to the structural and mechanical detriment of most metals. In an oxidizing flame, the inner cone acquires a purplish tinge, gets pinched and smaller at the tip, and the sound of the flame gets harsh. A slightly oxidizing flame is used in braze-welding and bronze-surfacing while a more strongly oxidizing flame is used in fusion welding certain brasses and bronzes

The size of the flame can be adjusted to a limited extent by the valves on the torch and by the regulator settings, but in the main it depends on the size of the orifice in the tip. In fact, the tip should be chosen first according to the job at hand, and then the regulators set accordingly.

Welding

The flame is applied to the base metal and held until a small puddle of molten metal is formed. The puddle is moved along the path where the weld bead is desired. Usually, more metal is added to the puddle as it is moved along by dipping metal from a welding rod or filler rod into the molten metal puddle. The metal puddle will travel towards where the metal is the hottest. This is accomplished through torch manipulation by the welder.

The amount of heat applied to the metal is a function of the welding tip size, the speed of travel, and the welding position. The flame size is determined by the welding tip size. The proper tip size is determined by the metal thickness and the joint design.

Welding gas pressures using oxy-acetylene are set in accordance with the manufacturer's recommendations. The welder will modify the speed of welding travel to maintain a uniform bead width. Uniformity is a quality attribute indicating good workmanship. Trained welders are taught to keep the bead the same size at the beginning of the weld as at the end. If the bead gets too wide, the welder increases the speed of welding travel. If the bead gets too narrow or if the weld puddle is lost, the welder slows down the speed of travel. Welding in the vertical or overhead positions is typically slower than welding in the flat or horizontal positions.

The welder must add the filler rod to the molten puddle. The welder must also keep the filler metal in the hot outer flame zone when not adding it to the puddle to protect filler metal from oxidation. Do not let the welding flame burn off the filler metal. The metal will not wet into the base metal and will look like a series of cold dots on the base metal. There is very little strength in a cold weld. When the filler metal is properly added to the molten puddle, the resulting weld will be stronger than the original base metal.

Welding lead or 'lead burning' was much more common in the 19th century to make some pipe connections and tanks. Great skill is required but can be quickly learned. In building construction today some lead flashing is welded but soldered copper flashing is much more common in America. In the automotive body collision industry before the 1980s, oxyacetylene gas torch welding was seldom used to weld sheetmetal, since warpage was a byproduct besides the excess heat. Automotive body repair methods at the time were crude and yielded improprieties until MIG welding became the industry standard. Since the 1970s, when high strength steel became the standard for automotive manufacturing, electric welding became the preferred method. After the 1980s, the oxyacetylene torch fell out of use for sheetmetal welding in the industrialized world.

Cutting

For cutting, the setup is a little different. A cutting torch has a 60- or 90-degree angled head with orifices placed around a central jet. The outer jets are for preheat flames of oxygen and acetylene. The central jet carries only oxygen for cutting. The use of several preheating flames rather than a single flame makes it possible to change the direction of the cut as desired without changing the position of the nozzle or the angle which the torch makes with the direction of the cut, as well as giving a better preheat balance. Manufacturers have developed custom tips for Mapp, propane, and polypropylene gases to optimize the flames from these alternate fuel gases.

The flame is not intended to melt the metal, but to bring it to its ignition temperature.

The torch's trigger blows extra oxygen at higher pressures down the torch's third tube out of the central jet into the workpiece, causing the metal to burn and blowing the resulting molten oxide through to the other side. The ideal kerf is a narrow gap with a sharp edge on either side of the workpiece; overheating the workpiece and thus melting through it causes a rounded edge.

Oxygen Rich Butane Torch Flame

Fuel Rich Butane Torch Flame

Cutting a rail just before renewing the rails and the ballast.

Cutting is initiated by heating the edge or leading face (as in cutting shapes such as round rod) of the steel to the ignition temperature (approximately bright cherry red heat) using the pre-heat jets only, then using the separate cutting oxygen valve to release the oxygen from the central jet.

The oxygen chemically combines with the iron in the ferrous material to oxidize the iron quickly into molten iron oxide, producing the cut. Initiating a cut in the middle of a workpiece is known as piercing.

It is worth noting several things at this point:

- The oxygen flowrate is critical; too little will make a slow ragged cut, while too much will waste oxygen and produce a wide concave cut. Oxygen lances and other custom made torches do not have a separate pressure control for the cutting oxygen, so the cutting oxygen pressure must be controlled using the oxygen regulator. The oxygen cutting pressure should match the cutting tip oxygen orifice. Consult the tip manufacturer's equipment data for the proper cutting oxygen pressures for the specific cutting tip.

- The oxidation of iron by this method is highly exothermic. Once it has started, steel can be cut at a surprising rate, far faster than if it were merely melted through. At this point, the pre-heat jets are there purely for assistance. The rise in temperature will be obvious by the intense glare from the ejected material, even through proper goggles. (*A thermic lance is a tool that also uses rapid oxidation of iron to cut through almost any material.*)

- Since the melted metal flows out of the workpiece, there must be room on the opposite side of the workpiece for the spray to exit. When possible, pieces of metal are cut on a grate that lets the melted metal fall freely to the ground. The same equipment can be used for oxyacetylene blowtorches and welding torches, by exchanging the part of the torch in front of the torch valves.

For a basic oxy-acetylene rig, the cutting speed in light steel section will usually be nearly twice as fast as a petrol-driven cut-off grinder. The advantages when cutting large sections are obvious: an oxy-fuel torch is light, small and quiet and needs very little effort to use, whereas a cut-off grinder is heavy and noisy and needs considerable operator exertion and may vibrate severely, leading to stiff hands and possible long-term vibration white finger. Oxy-acetylene torches can easily cut through ferrous materials in excess of 200 mm (8 inches). Oxygen lances are used in scrapping operations and cut sections thicker than 200 mm (8 inches). Cut-off grinders are useless for these kinds of application.

Robotic oxy-fuel cutters sometimes use a high-speed divergent nozzle. This uses an oxygen jet that opens slightly along its passage. This allows the compressed oxygen to expand as it leaves, forming a high-velocity jet that spreads less than a parallel-bore nozzle, allowing a cleaner cut. These are not used for cutting by hand since they need very accurate positioning above the work. Their ability to produce almost any shape from large steel plates gives them a secure future in shipbuilding and in many other industries.

Oxy-propane torches are usually used for cutting up scrap to save money, as LPG is far cheaper joule for joule than acetylene, although propane does not produce acetylene's very neat cut profile. Propane also finds a place in production, for cutting very large sections.

Oxy-acetylene can cut only low- to medium-carbon steels and wrought iron. High-carbon steels are difficult to cut because the melting point of the slag is closer to the melting point of the parent metal, so that the slag from the cutting action does not eject as sparks but rather mixes with the clean melt near the cut. This keeps the oxygen from reaching the clean metal and burning it. In the case of cast iron, graphite between the grains and the shape of the grains themselves interfere with the cutting action of the torch. Stainless steels cannot be cut either because the material does not burn readily.

Safety

Oxygas welding station (keep cylinders and hoses away from the flame)

Gas welding/cutting goggles and safety helmet

Oxyacetylene welding/cutting is not difficult, but there are a good number of subtle safety points that should be learned such as:

- More than 1/7 the capacity of the cylinder should not be used per hour. This causes the acetone inside the acetylene cylinder to come out of the cylinder and contaminate the hose and possibly the torch.

- Acetylene is dangerous above 1 atm (15 psi) pressure. It is unstable and explosively decomposes.

- Proper ventilation when welding will help to avoid large chemical exposure.

The Importance of Eye Protection

Proper protection such as welding goggles should be worn at all times, including to protect the eyes against glare and flying sparks. Special safety eyewear must be used—both to protect the welder and to provide a clear view through the yellow-orange flare given off by the incandescing flux. In the 1940s cobalt melters' glasses were borrowed from steel foundries and were still available until the 1980s. However, the lack of protection from impact, ultra-violet, infrared and blue light caused severe eyestrain and eye damage. Didymium eyewear, developed for glassblowers in the 1960s, was also borrowed—until many complained of eye problems from excessive infrared, blue light, and insufficient shading. Today very good eye protection can be found designed especially for gas-welding aluminum that cuts the sodium orange flare completely and provides the necessary protection from ultraviolet, infrared, blue light and impact, according to ANSI Z87-1989 safety standards for a Special Purpose Lens.

Fuel Leakage

Fuel gases that are denser than air (Propane, Propylene, MAPP, Butane, etc...), may collect in low areas if allowed to escape. To avoid an ignition hazard, special care should be taken when using these gases over areas such as basements, sinks, storm drains, etc. In addition, leaking fittings may catch fire during use and pose a risk to personnel as well as property.

Safety With Cylinders

When using fuel and oxygen tanks they should be fastened securely upright to a wall or a post or a portable cart. An oxygen tank is especially dangerous for the reason that the oxygen is at a pressure of 21 MPa (3000 lbf/in² = 200 atmospheres) when full, and if the tank falls over and its valve strikes something and is knocked off, the tank will effectively become an extremely deadly flying missile propelled by the compressed oxygen, capable of even breaking through a brick wall. For this reason, never move an oxygen tank around without its valve cap screwed in place.

On an oxyacetylene torch system there will be three types of valves, the tank valve, the regulator valve, and the torch valve. There will be a set of these three valves for each gas. The gas in the tanks or cylinders is at high pressure. Oxygen cylinders are generally filled to approximately 2200 psi. The regulator converts the high pressure gas to a low pressure stream suitable for welding. Acetylene cylinders must be maintained in an upright position to prevent the internal acetone and acetylene from separating in the filler material.

Chemical Exposure

A less obvious hazard of welding is exposure to harmful chemicals. Exposure to certain metals, metal oxides, or carbon monoxide can often lead to severe medical conditions. Damaging chemicals can be produced from the fuel, from the work-piece, or from a protective coating on the work-piece. By increasing ventilation around the welding environment, the welders will have much less exposure to harmful chemicals from any source.

The most common fuel used in welding is acetylene, which has a two-stage reaction. The primary chemical reaction involves the acetylene disassociating in the presence of oxygen to produce heat, carbon monoxide, and hydrogen gas: $C_2H_2 + O_2 \rightarrow 2CO + H_2$. A secondary reaction follows where the carbon monoxide and hydrogen combine with more oxygen to produce carbon dioxide and water vapor. When the secondary reaction does not burn all of the reactants from the primary reaction, the welding process can produce large amounts of carbon monoxide, and it often does. Carbon monoxide is also the byproduct of many other incomplete fuel reactions.

Almost every piece of metal is an alloy of one type or another. Copper, aluminium, and other base metals are occasionally alloyed with beryllium, which is a highly toxic metal. When a metal like this is welded or cut, high concentrations of toxic beryllium fumes are released. Long-term exposure to beryllium may result in shortness of breath, chronic cough, and significant weight loss, accompanied by fatigue and general weakness. Other alloying elements such as arsenic, manganese, silver, and aluminium can cause sickness to those who are exposed.

More common are the anti-rust coatings on many manufactured metal components. Zinc, cadmium, and fluorides are often used to protect irons and steels from oxidizing. Galvanized metals have a very heavy zinc coating. Exposure to zinc oxide fumes can lead to a sickness named "metal fume fever". This condition rarely lasts longer than 24 hours, but severe cases can be fatal. Not unlike common influenza, fevers, chills, nausea, cough, and fatigue are common effects of high zinc oxide exposure.

Flashback

Flashback is the condition of the flame propagating down the hoses of an oxy-fuel welding and cutting system. To prevent such a situation a flashback arrestor is usually employed. The flame burns backwards into the hose, causing a popping or squealing noise. It can cause an explosion in the hose with the potential to injure or kill the operator. Using a lower pressure than recommended can cause a flashback.

Electric Resistance Welding

Electric resistance welding (ERW) refers to a group of welding process such as spot and seam welding that produce coalescence of faying surfaces where heat to form the weld is generated by the electrical resistance of material combined with the time and the force used to hold the materials together during welding. Some factors influencing heat or welding temperatures are the proportions of the workpieces, the metal coating or the lack of coating, the electrode materials, electrode geometry, electrode pressing force, electrical current and length of welding time. Small pools of molten metal are formed at the point of most electrical resistance (the connecting or "faying" surfaces) as an electrical current (100–100,000 A) is passed through the metal. In general, resistance welding methods are efficient and cause little pollution, but their applications are limited to relatively thin materials and the equipment cost can be high (although in production situations the cost per weld may be low).

Spot Welding

Spot welding is a resistance welding method used to join two or more overlapping metal sheets, studs, projections, electrical wiring hangers, some heat exchanger fins, and some tubing. Usually power sources and welding equipment are sized to the specific thickness and material being welded together. The thickness is limited by the output of the welding power source and thus the equipment range due to the current required for each application. Care is taken to eliminate contaminants between the faying surfaces. Usually, two copper electrodes are simultaneously used to clamp the metal sheets together and to pass current through the sheets. When the current is passed through the electrodes to the sheets, heat is generated due to the higher electrical resistance where the surfaces contact each other. As the electrical resistance of the material causes a heat buildup in the work pieces between the copper electrodes, the rising temperature causes a rising resistance, and results in a molten pool contained most of the time between the electrodes. As the heat dissipates throughout the workpiece in less than a second (resistance welding time is generally programmed as a quantity of AC cycles or milliseconds) the molten or plastic state grows to meet the welding tips. When the current is stopped the copper tips cool the spot weld, causing the metal to solidify under pressure. The water cooled copper electrodes remove the surface heat quickly, accelerating the solidification of the metal, since copper is an excellent conductor. Resistance spot welding typically employs electrical power in the form of direct current, alternating current, medium frequency half-wave direct current, or high-frequency half wave direct current.

Spot welder

If excessive heat is applied or applied too quickly, or if the force between the base materials is too low, or the coating is too thick or too conductive, then the molten area may extend to the exterior of the work pieces, escaping the containment force of the electrodes (often up to 30,000 psi). This burst of molten metal is called expulsion, and when this occurs the metal will be thinner and have less strength than a weld with no expulsion. The common method of checking a weld's quality is a peel test. An alternative test is the restrained tensile test, which is much more difficult to perform, and requires calibrated equipment. Because both tests are destructive in nature (resulting in the loss of salable material), non-destructive methods such as ultrasound evaluation are in various states of early adoption by many OEMs.

The advantages of the method include efficient energy use, limited workpiece deformation, high production rates, easy automation, and no required filler materials. When high strength in shear is needed, spot welding is used in preference to more costly mechanical fastening, such as riveting. While the shear strength of each weld is high, the fact that the weld spots do not form a continuous seam means that the overall strength is often significantly lower than with other welding methods, limiting the usefulness of the process. It is used extensively in the automotive industry— cars can have several thousand spot welds. A specialized process, called shot welding, can be used to spot weld stainless steel.

There are three basic types of resistance welding bonds: solid state, fusion, and reflow braze. In a *solid state bond*, also called a thermo-compression bond, dissimilar materials with dissimilar grain structure, e.g. molybdenum to tungsten, are joined using a very short heating time, high weld energy, and high force. There is little melting and minimum grain growth, but a definite bond and grain interface. Thus the materials actually bond while still in the solid state. The bonded materials typically exhibit excellent shear and tensile strength, but poor peel strength. In a *fusion bond*, either similar or dissimilar materials with similar grain structures are heated to the melting point (liquid state) of both. The subsequent cooling and combination of the materials forms a "nugget" alloy of the two materials with larger grain growth. Typically, high weld energies at either short or long weld times, depending on physical characteristics, are used to produce fusion bonds. The bonded materials usually exhibit excellent tensile, peel and shear strengths. In a *reflow braze bond*, a resistance heating of a low temperature brazing material, such as gold or solder, is used to join either dissimilar materials or widely varied thick/thin material combinations. The brazing material must "wet" to each part and possess a lower melting point than the two workpieces. The resultant bond has definite interfaces with minimum grain growth. Typically the process requires a longer (2 to 100 ms) heating time at low weld energy. The resultant bond exhibits excellent tensile strength, but poor peel and shear strength.

Seam Welding

Resistance seam welding is a process that produces a weld at the faying surfaces of two similar metals. The seam may be a butt joint or an overlap joint and is usually an automated process. It differs from butt welding in that butt welding typically welds the entire joint at once and seam welding forms the weld progressively, starting at one end. Like spot welding, seam welding relies on two electrodes, usually made from copper, to apply pressure and current. The electrodes are disc shaped and rotate as the material passes between them. This allows the electrodes to stay in constant contact with the material to make long continuous welds. The electrodes may also move or assist the movement of the material.

A transformer supplies energy to the weld joint in the form of low voltage, high current AC power. The joint of the work piece has high electrical resistance relative to the rest of the circuit and is heated to its melting point by the current. The semi-molten surfaces are pressed together by the welding pressure that creates a fusion bond, resulting in a uniformly welded structure. Most seam welders use water cooling through the electrode, transformer and controller assemblies due to the heat generated. Seam welding produces an extremely durable weld because the joint is forged due to the heat and pressure applied. A properly welded joint formed by resistance welding is typically stronger than the material from which it is formed.

A common use of seam welding is during the manufacture of round or rectangular steel tubing. Seam welding has been used to manufacture steel beverage cans but is no longer used for this as modern beverage cans are seamless aluminum.

There are two modes for seam welding: Intermittent and continuous. In intermittent seam welding, the wheels advance to the desired position and stop to make each weld. This process continues until the desired length of the weld is reached. In continuous seam welding, the wheels continue to roll as each weld is made.

Low-frequency Electric Resistance Welding

Low-frequency electric resistance welding, LF-ERW, is an obsolete method of welding seams in oil and gas pipelines. It was phased out in the 1970s but as of 2015 some pipelines built with this method remained in service.

Electric resistance welded (ERW) pipe is manufactured by cold-forming a sheet of steel into a cylindrical shape. Current is then passed between the two edges of the steel to heat the steel to a point at which the edges are forced together to form a bond without the use of welding filler material. Initially this manufacturing process used low frequency A.C. current to heat the edges. This low frequency process was used from the 1920s until 1970. In 1970, the low frequency process was superseded by a high frequency ERW process which produced a higher quality weld.

Over time, the welds of low frequency ERW pipe was found to be susceptible to selective seam corrosion, hook cracks, and inadequate bonding of the seams, so low frequency ERW is no longer used to manufacture pipe. The high frequency process is still being used to manufacture pipe for use in new pipeline construction.

Other Methods

Other ERW methods include flash welding, resistance projection welding, and upset welding.

Fillet Weld

Fillet welding refers to the process of joining two pieces of metal together whether they be perpendicular or at an angle. These welds are commonly referred to as Tee joints which are two pieces of metal perpendicular to each other or Lap joints which are two pieces of metal that overlap and are welded at the edges. The weld is aesthetically triangular in shape and may have a concave, flat or convex surface depending on the welder's technique. Welders use fillet welds when connecting flanges to pipes, welding cross sections of infrastructure, and when fastening metal by bolts isn't strong enough.

Making a fillet weld with gas metal arc welding

Aspects

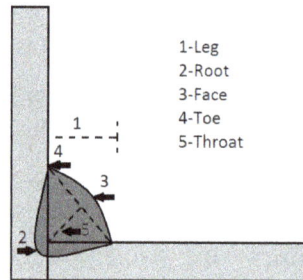

Parts of a fillet weld

There are 5 pieces to each Fillet weld known as the Root, Toe, Face, Leg and Throat. The root of the weld is the part of deepest penetration which is the opposite angle of the Hypotenuse. The toes of the weld are essentially the edges or the points of the hypotenuse. The face of the weld is the outer visual or hypotenuse that you see when looking at a fillet weld. The legs are the opposite and adjacent sides to the triangular fillet weld. The leg length is usually designated as the size of the weld. The throat of the weld is the distance from the center of the face to the root of the weld. Typically the depth of the throat should be at least as thick as the thickness of metal you are welding.

Notation

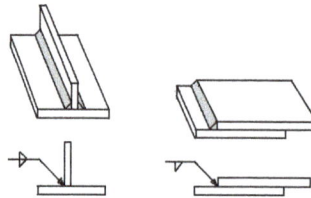

Fillet weld notation

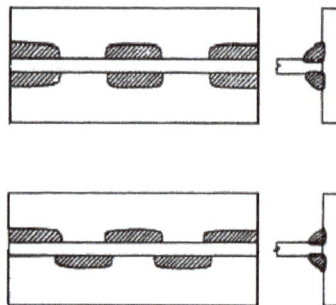

Intermittent fillet welds

Fillet welding notation is important to recognize when reading blueprints. The use of this notation tells the welder exactly what is expected from the manufacturer. The symbol for a fillet weld is in the shape of a triangle. This triangle will lay either below a flat line or above it with an arrow coming off of the flat line pointing to a joint. The flat line is called "reference line". The side on which the triangle symbol is placed is important because it gives an indication on which side of the joint is to be intersected by the weld. It is recognized that there are two different approaches in the global market to designate the arrow side and other side on drawings; a description of the two approaches is contained in International Standard ISO 2553, they are called "A-System" (which is more commonly used in Europe) and "B-System" (which is basically the ANSI/AWS system used in the US). In "A-System" two parallel lines are used as reference line: one is a continuous line, the other is a dashed line. In the "B-System", there is only one reference line, which is a continuous line. If there is a single reference line (B-System) and the triangle is positioned below the line, then the weld is going to be on the arrow side. If there is a single reference line ("B-System") and the triangle is positioned above the line, then the weld is going to be on the opposite side of the arrow. When you find an arrow pointing to a joint with two triangles, one sitting below and one sitting above the line even with each other, then there is intended to be a fillet weld on the arrow side of the joint as well as the opposite side of the joint. If the weld is to be continuous around a piece of metal such as a pipe or square, then a small circle will be around the point where the flat line and arrow pointing to the joint are connected. Manufacturers also include the strength that the weld must be. This is indicated by a letter and number combination just before the flat line. Examples of this are "E70" meaning the arc electrode must have a tensile strength of 70,000 pounds-force per square inch (480,000 kPa; 4,900 kgf/cm^2).There are also symbols that describe the aesthetics of the weld. A gentle curve pointing away from the hypotenuse means a concave weld is required, a straight line parallel with the hypotenuse calls for a flat faced weld, and a gentle curve towards the hypotenuse calls for a convex weld. The surface of the weld can be manipulated either by welding technique or by use of machining or grinding tools after the weld is completed. When reading a manufacturers blueprints, you might also come across weld dimensions. The weld can be sized in many different ways such as the length of the weld, the measurements of the legs of the weld, and the spaces between welds. Along with a triangle, there will usually be a size for the weld for example ($\frac{1}{8}$"x$\frac{3}{8}$") to the left of the triangle. This means that the vertical leg of the weld is to be $\frac{1}{8}$" whereas the horizontal leg is to $\frac{3}{8}$". To the right of the triangle, there will be a measurement of exactly how long the weld is supposed to be. If the measurements of the drawing are in mm the welds are likewise measured in mm. For example, the weld would be 3 x 10, the mm being understood automatically.

Intermittent Fillet Welds

An intermittent fillet weld is one that is not continuous across a joint. These welds are portrayed as a set of two numbers to the right of the triangle instead of just one. The first number as mentioned earlier refers to the length of the weld. The second number, separated from the first by a "-", refers to the pitch. The pitch is a measurement from midpoint to midpoint of the intermittent welds. Intermittent welding is used when either a continuous weld is not necessary, or when a continuous weld threatens the joint by warping. In some cases intermittent welds are staggered on both sides of the joint. In this case, the notation of the two triangles aren't directly on top of each other. Instead, the side of the joint to receive the first weld will have a triangle further to the left than the following side's triangle notation. As an end result of alternating intermittent fillet welds at each side, the space between welds on one side of the joint will be the midpoint of the opposite side's weld.

Plasma Cutting

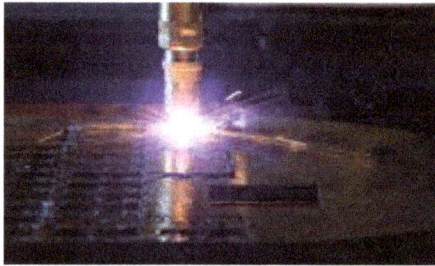

CNC Plasma Cutting

Plasma cutting performed by an industrial robot

Plasma cutting is a process that cuts through electrically conductive materials by means of an accelerated jet of hot plasma. Typical materials cut with a plasma torch include steel, aluminum, brass and copper, although other conductive metals may be cut as well. Plasma cutting is often used in fabrication shops, automotive repair and restoration, industrial construction, and salvage and scrapping operations. Due to the high speed and precision cuts combined with low cost, plasma cutting sees widespread use from large-scale industrial CNC applications down to small hobbyist shops.

Process

The basic plasma cutting process involves creating an electrical channel of superheated, electrically ionized gas i.e. plasma from the plasma cutter itself, through the work piece to be cut, thus forming a completed electric circuit back to the plasma cutter via a grounding clamp. This is accomplished by a compressed gas (oxygen, air, inert and others depending on material being cut) which is blown through a focused nozzle at high speed toward the work piece. An electrical arc is then formed within the gas, between an electrode near or integrated into the gas nozzle and the work piece itself. The electrical arc ionizes some of the gas, thereby creating an electrically conductive channel of plasma. As electricity from the cutter torch travels down this plasma it delivers sufficient heat to melt through the work piece. At the same time, much of the high velocity plasma and compressed gas blow the hot molten metal away, thereby separating i.e. cutting through the work piece.

Freehand cut of a thick steel plate

Plasma cutting is an effective means of cutting thin and thick materials alike. Hand-held torches can usually cut up to 38mm thick steel plate, and stronger computer-controlled torches can cut steel up to 150 mm thick. Since plasma cutters produce a very hot and very localized "cone" to cut with, they are extremely useful for cutting sheet metal in curved or angled shapes.

History

Plasma cutting with a tilting head

Plasma cutting grew out of plasma welding in the 1960s, and emerged as a very productive way to cut sheet metal and plate in the 1980s. It had the advantages over traditional "metal against metal" cutting of producing no metal chips, giving accurate cuts, and producing a cleaner edge than oxy-fuel cutting. Early plasma cutters were large, somewhat slow and expensive and, therefore, tended to be dedicated to repeating cutting patterns in a "mass production" mode.

As with other machine tools, CNC (computer numerical control) technology was applied to plasma cutting machines in the late 1980s into the 1990s, giving plasma cutting machines greater flexibility to cut diverse shapes "on demand" based on a set of instructions that were programmed into the machine's numerical control. These CNC plasma cutting machines were, however, generally limited to cutting patterns and parts in flat sheets of steel, using only two axes of motion (referred to as X Y cutting).

Safety

Proper eye protection and face shields are needed to prevent eye damage called arc eye as well as damage from debris. It is recommended to use green lens shade #5. OSHA recommends a shade 8 for arc current less than 300, but notes that *"These values apply where the actual arc is clearly seen. Experience has shown that lighter filters may be used when the arc is hidden by the workpiece."* Lincoln Electric, a manufacturer of plasma cutting equipment, says, *"Typically a darkness shade of #7 to #9 is acceptable."* Longevity Global, Inc., another manufacturer, offers this more specific table for Eye Protection for Plasma Arc Cutting at lower amperages :

Current Level in Amps	Minimum Shade Number
Below 20	#4
20-40	#5
40-60	#6
60-80	#8

Leather gloves, apron and jacket are also recommended to prevent burns from sparks and debris.

Starting Methods

Plasma cutters use a number of methods to start the arc. In some units, the arc is created by putting the torch in contact with the work piece. Some cutters use a high voltage, high frequency circuit to start the arc. This method has a number of disadvantages, including risk of electrocution, difficulty of repair, spark gap maintenance, and the large amount of radio frequency emissions. Plasma cutters working near sensitive electronics, such as CNC hardware or computers, start the pilot arc by other means. The nozzle and electrode are in contact. The nozzle is the cathode, and the electrode is the anode. When the plasma gas begins to flow, the nozzle is blown forward. A third, less common method is capacitive discharge into the primary circuit via a silicon controlled rectifier.

Inverter Plasma Cutters

Plasma cutting

Analog plasma cutters, typically requiring more than 2 kilowatts, use a heavy mains-frequency transformer. Inverter plasma cutters rectify the mains supply to DC, which is fed into a high-frequency transistor inverter between 10 kHz to about 200 kHz. Higher switching frequencies allow smaller transformer resulting in overall size and weight reduction.

The transistors used were initially MOSFETs, but are now increasingly using IGBTs. With paralleled MOSFETs, if one of the transistors activates prematurely it can lead to a cascading failure of one quarter of the inverter. A later invention, IGBTs, are not as subject to this failure mode. IGBTs can be generally found in high current machines where it is not possible to parallel sufficient MOSFET transistors.

The switch mode topology is referred to as a dual transistor off-line forward converter. Although lighter and more powerful, some inverter plasma cutters, especially those without power factor correction, cannot be run from a generator (that means manufacturer of the inverter unit forbids doing so; it is only valid for small, light portable generators). However newer models have internal circuitry that allow units without power factor correction to run on light power generators.

CNC Cutting Methods

Some plasma cutter manufacturers build CNC cutting tables, and some have the cutter built into the table. CNC tables allow a computer to control the torch head producing clean sharp cuts. Modern CNC plasma equipment is capable of multi-axis cutting of thick material, allowing opportunities for complex welding seams that are not possible otherwise. For thinner material, plasma cutting is being progressively replaced by laser cutting, due mainly to the laser cutter's superior hole-cutting abilities.

A specialized use of CNC Plasma Cutters has been in the HVAC industry. Software processes information on ductwork and creates flat patterns to be cut on the cutting table by the plasma torch. This technology has enormously increased productivity within the industry since its introduction in the early 1980s.

CNC Plasma Cutters are also used in many workshops to create decorative metalwork. For instance, commercial and residential signage, wall art, address signs, and outdoor garden art.

In recent years there has been even more development. Traditionally the machines' cutting tables were horizontal, but now vertical CNC plasma cutting machines are available, providing for a smaller footprint, increased flexibility, optimum safety and faster operation.

CNC Plasma Cutting Configurations

There are 3 main configurations of CNC Plasma Cutting, and they are largely differentiated by the forms of materials before processing, and the flexibility of the cutting head.

2 Dimensional / 2-Axis Plasma Cutting

This is the most common and conventional form of CNC Plasma Cutting. Producing flat profiles, where the cut edges are at 90 Degrees to the material surface. High powered cnc plasma cutting beds are configured in this way, able to cut profiles from metal plate up to 150mm thick.

3 Dimensional / 3+ Axis Plasma Cutting

Once again, a process for producing flat profiles from sheet or plate metal, however with the introduction of an additional axis of rotation, the cutting head of a CNC Plasma Cutting machine can tilt whilst being taken through a conventional 2 dimensional cutting path. The result of this is cut edges at an angle other than 90 Degrees to the material surface, for example 30-45 Degree angles. This angle is continuous throughout the thickness of the material. This is typically applied in situations where the profile being cut is to be used as part of a welded fabrication as the angled edge forms part of the weld preparation. When the weld preparation is applied during the cnc plasma cutting process, secondary operations such as grinding or machining can be avoided, reducing cost. The angular cutting capability of 3 Dimensional plasma cutting can also be used to create countersunk holes and chamfer edges of profiled holes.

Tube & Section Plasma Cutting

Used in the processing of tube, pipe or any form of long section. The plasma cutting head usually remains stationary whilst the workpiece is fed through, and rotated around its longitudinal axis. There are some configurations where, as with 3 Dimensional Plasma Cutting, the cutting head can tilt and rotate. This allows angled cuts to be made through the thickness of the tube or section, commonly taken advantage of in the fabrication of process pipework where cut pipe can be provided with a weld preparation in place of a straight edge.

New Technology

In the past decade plasma torch manufacturers have engineered new models with a smaller nozzle and a thinner plasma arc. This allows near-laser precision on plasma cut edges. Several manufacturers have combined precision CNC control with these torches to allow fabricators to produce parts that require little or no finishing.

Costs

Plasma torches were once quite expensive. For this reason they were usually only found in professional welding shops and very well-stocked private garages and shops. However, modern plasma torches are becoming cheaper, and now are within the price range of many hobbyists. Older units may be very heavy, but still portable, while some newer ones with inverter technology weigh only a little, yet equal or exceed the capacities of older ones.

Fusion Welding

Fusion welding is a generic term for welding processes that rely upon melting to join materials of similar compositions and melting points. Due to the high-temperature phase transitions inherent to these processes, a heat-affected zone is created in the material (although some techniques, like beam welding, often minimize this effect by introducing comparatively little heat into the work-piece).

Classification of fusion welding processes based on energy source, thermal source, mechanical loading and shielding

In contrast to fusion welding, solid-state welding does not involve the melting of materials.

Applications

Fusion welding was a key factor in the creation of modern civilization. It is also very important for maintaining civilization. Fusion welding holds a key role in construction. Besides bolts and rivets there are no other practical method for joining pieces of metal securely. Fusion welding is used to make many everyday items. From construction to airplanes to cars fusion welding is very common. Another common use of fusion welding is in artwork. There is a large community that uses both arc and flame contact welding artwork.

Types

Electrical

Arc

Arc welding is one of the many types of fusion welding. Arc welding joins two pieces of metal together by using an intermediate filler metal. The way this works is by completing an electrical circuit to create an electrical arc. This electrical arc is 6500°F in its center. This electrical arc is created at the tip of the filler metal. As the arc melts the metal it is moved either by a person or a machine along the gap in the metals creating a bond. This method is very common as it is typically done with a hand held machine. Arc welding machines are portable and can be brought on to job sites and in hard to reach areas. Arc welding is also the most common method of underwater welding. Electrical arcs form between points separated by a gas. In the process of underwater welding a bubble of gas is blown around the area being welded so that an electrical arc may form. Underwater welding has many applications. Ship hulls are repaired and oil rigs are maintained with underwater arc welding.

Resistance welding is done using two electrodes. Each electrode comes into contact with one of the pieces being welded. The two pieces of metal are then pressed together between the electrodes.

While they are being pressed an electric current is ran through them. As this happens the pieces of metal begin to heat up at the point where they come into contact. The current is passed through the metal until it is hot enough to the point where the two pieces melt and conjoin. As the metal cools the bond is solidified. This process requires large amounts of electricity. In most cases transformers are needed to provide enough Amps. Resistance welding is a very prevalent form of fusion welding. Resistance welding is used in the manufacturing of automobiles and construction equipment.

Laser Beam Welding

Conduction welding also known as laser welding or radiation welding is a highly precise form of fusion welding. Laser is an acronym for Light Amplification by Stimulated Emission of Radiation. The laser emits light in bursts called pumps. These bursts are aimed at the seam of the metals desired to be conjoined. As the laser bursts it is guided along the seam. These intense bursts melt the metal. The two metals when melted mix with each other. Once it has cooled the seam created is a strong bond. Lasers are efficient because they can be configured to make multiple welds at once. The laser beam can be split and sent to multiple locations greatly reducing the cost and amount of energy required. Laser beam welding finds applications in the automotive industry.

Induction

Induction welding is a form of resistance welding. In induction welding however there are no points of contact between the metal being welding and the electrical source or the welder. In induction welding a coil is wrapped around a cylinder. This coil causes a magnetic field across the surface of the metal inside. This magnetic field flows in the opposite direction of the magnetic field on the inside of the cylinder. These magnetic flows impede each other. This heats up the metal and causes either the damaged or separate edges to melt together.

Chemical

Oxyfuel

Flame contact welding is a very common form of welding. The most popular kind of flame contact welding is oxyfuel gas welding. Flame contact welding uses a flame exposed to the surfaced of the metals being welding to melt and then join them. Oxyacetylene is the most common form of oxyfuel welding. Oxyfuel uses oxygen as a primary ignition source in tandem with another gas such as acetylene to produce a flame which is 2500°C at the tip and 2800-3500°C at the tip of the inner cone. Other gasses such as propane and methanol can be used for oxyfuel welding. Acetylene is the most common gas used in oxyfuel welding.

Solid Reactant

Solid reactant welding uses reactions between elements and compounds. Certain compounds when mixed create an exothermic chemical reaction, meaning they give off heat. A very common reaction uses is thermite. Thermite is a combination of a metal oxide (rust) and aluminum. This reaction produces heat over 4000°F. Solid reactant compounds are channeled to the two pieces of metal being joined. Once in place a catalyst is used to start the reaction. This catalyst can be a chemical or another heat source. The heat created melts the metals being joined. Once it cools a bond is formed. Solid reactant has been used throughout history. From welding together train tracks to entering banks vaults it has many niche uses.

Magnetic Pulse Welding

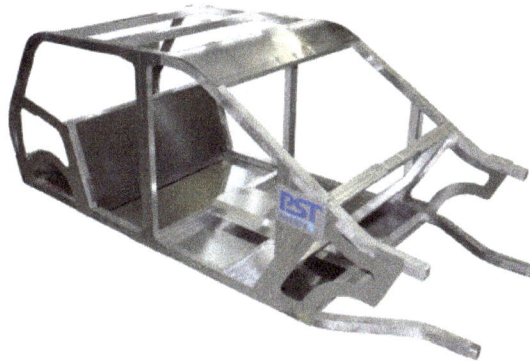

Magnetic pulse welded space frame

Magnetic pulse welding (MPW) is a solid state welding process that uses magnetic forces to weld two workpieces together. The welding mechanism is most similar to that of explosion welding. Magnetic pulse welding started in the early 1970s, when the automotive industry began to use solid state welding. The biggest advantage using magnetic pulse welding is that the formation of brittle intermetallic phases is avoided. Therefore, dissimilar metals can be welded, which cannot be joined by fusion welding. With magnetic pulse welding high quality welds in similar and dissimilar metals can be made in microseconds without the need for shielding gases or welding consumables.

Process

Magnetic pulse welded HVAC pressure vessel

Magnetic pulse welding is based on a very short electromagnetic pulse (<100μs), which is obtained by a fast discharge of capacitors through low inductance switches into a coil. The pulsed current with a very high amplitude and frequency (500kA and 15 kHz) produces a high-density magnetic field, which creates an eddy current in one of the work pieces. Repulsive Lorentz forces are created and a high magnetic pressure well beyond the material yield strength causing acceleration and one of the work pieces impacts onto the other part with a collision velocity up to 500 m/s.

During magnetic pulse welding a high plastic deformation is developed along with high shear strain and oxide disruption thanks to the jet and high temperatures near the collision zone. This leads to solid state weld due to the microstructure refinement, dislocation cells, slip bends, micro twins and local recrystallization.

Principles

In order to get a strong weld, several conditions have to be reached:

- *Jetting condition:* the collision has to be subsonic compared to the local materials speed of sound to generate a jet.

- *High pressure regime:* the impact velocity has to be sufficient to obtain a hydrodynamic regime, otherwise the parts will only be crimped or formed.

- *No fusion during the collision:* If the pressure is too high, the materials can locally melt and re-solidify. This can cause a weak weld.

The main difference between magnetic pulse welding and explosive welding is that the collision angle and the velocity are almost constant during the explosive welding process, while in magnetic pulse welding they continuously vary.

Advantages of MPW

- Allows welding of designs which with other processes are challenging or not possible.

- High-speed pulse lasts from 10 to 100 μs, the only time limitation is loading and unloading and capacitor charge time.

- High repeatability.

- Suited to mass-production: typically 1-5 million welds per year.

- Dissimilar metals welding is possible.

- Weld with no heat-affected zone.

- No need for filler materials.

- Green process: no smoke, no radiation and no extraction equipment required.

- High quality clean interface.

- Mechanical strength of the joint is stronger than that of the parent material.

- High precision obtainable by adjustment of magnetic field.

- No distortion.

- Almost zero residual stresses.

- No corrosion development in the welding area.

Explosion Welding

Explosion welding (EXW) is a solid state (solid-phase) process where welding is accomplished by accelerating one of the components at extremely high velocity through the use of chemical explosives. This process is most commonly utilized to clad carbon steel plate with a thin layer of corrosion resistant material (e.g., stainless steel, nickel alloy, titanium, or zirconium). Due to the nature of this process, producible geometries are very limited. They must be simple. Typical geometries produced include plates, tubing and tubesheets.

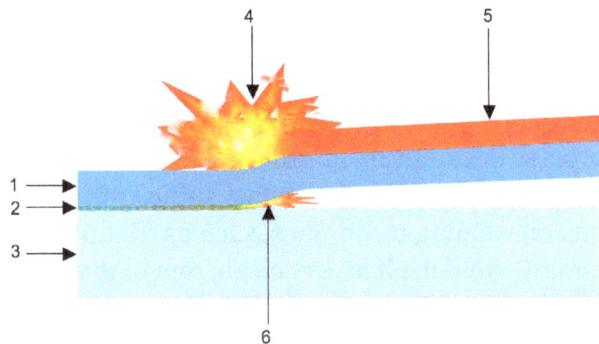

Explosion welding 1 Flyer (cladding). 2 Resolidified zone (needs to be minimised for welding of dissimilar materials). 3 Target (substrate). 4 Explosion. 5 Explosive powder. 6 Plasma jet.

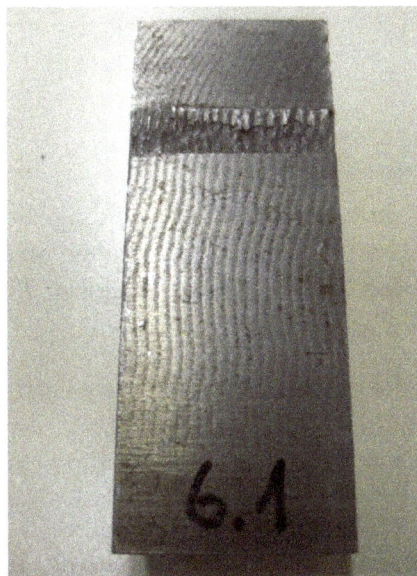

Polished section of an explosion weld with typical wave-structure

Development

Unlike other forms of welding such as arc welding (which was developed in the late 19th century), explosion welding was developed relatively recently, in the decades after World War II. Its origins, however, go back to World War I, when it was observed that pieces of shrapnel sticking to armor plating were not only embedding themselves, but were actually being welded to the metal. Since the extreme heat involved in other forms of welding did not play a role, it was concluded that the phenomenon was caused by the explosive forces acting on the shrapnel. These results were later duplicated in laboratory tests and, not long afterwards, the process was patented and put to use.

In 1962, DuPont applied for a patent on the explosion welding process, which was granted on June 23, 1964 under US Patent 3,137,937 and resulted in the use of the *Detaclad* trademark to describe the process. On July 22, 1996, Dynamic Materials Corporation completed the acquisition of Du-Pont's Detaclad operations for a purchase price of $5,321,850.

Recently, the response of inhomogeneous plates undergoing explosive welding was analytically modeled.

Advantages and Disadvantages

Explosion welding can produce a bond between two metals that cannot necessarily be welded by conventional means. The process does not melt either metal, instead plasticizing the surfaces of both metals, causing them to come into intimate contact sufficient to create a weld. This is a similar principle to other non-fusion welding techniques, such as friction welding. Large areas can be bonded extremely quickly and the weld itself is very clean, due to the fact that the surface material of both metals is violently expelled during the reaction.

A disadvantage of this method is that extensive knowledge of explosives is needed before the procedure may be attempted safely. Regulations for the use of high explosives may require special licensing.

Twin-carbon Arc Welding

Unlike single-carbon arc welding, in twin-carbon arc welding the arc is maintained between two carbon electrodes held in a special holder. Current is switched on and by operating the mechanism of arc length adjustment the two electrodes are brought closer. The two electrodes touch momentarily, then separate and thus an arc is established.

The size of the arc depends upon the distance between the electrode tips, electrode diameters and the welding current. The heat input to the job can be varied by changing the arc size or the distance between the arc and workpiece. After striking the arc, welding can be carried out in the same way as in TIG welding process.

An AC supply is recommended for twin-carbon arc welding. In case a DC supply is used, the positive electrode will disintegrate and consume at a much faster rate as compared to negative electrode, because two-thirds of the total heat is generated at the positive pole. This will produce an unstable arc and require frequent adjustment of the electrodes. In AC welding, because of alternate reversals of polarity, both the electrodes will be affected equally and present no problem.

The electrodes employed for twin-carbon arc welding are approximately of the same diameter as the workpiece thickness. The magnitude of arc current required for welding depends upon both electrode diameter and plate thickness. For example, an 8 mm diameter electrode will need about 65 amps to weld a mild steel sheet of thickness 3.5 mm and 80 Amps to weld a sheet of 6 mm thickness.

Twin-carbon arc welding, though more complex than single carbon arc welding, possesses the advantage that arc is independent of the job and can be moved anywhere without getting extinguished. Moreover, the workpiece is not a part of the electrical circuit.

Electrogas Welding

Electrogas welding (EGW) is a continuous vertical position arc welding process developed in 1961, in which an arc is struck between a consumable electrode and the workpiece. A shielding gas is sometimes used, but pressure is not applied. A major difference between EGW and its cousin electroslag welding is that the arc in EGW is not extinguished, instead remains struck throughout the welding process. It is used to make square-groove welds for butt and t-joints, especially in the shipbuilding industry and in the construction of storage tanks.

Operation

In EGW, the heat of the welding arc causes the electrode and workpieces to melt and flow into the cavity between the parts being welded. This molten metal solidifies from the bottom up, joining the parts being welded together. The weld area is protected from atmospheric contamination by a separate shielding gas, or by the gas produced by the disintegration of a flux-cored electrode wire. The electrode is guided into the weld area by either a consumable electrode guide tube, like the one used in electroslag welding, or a moving head. When the consumable guide tube is used, the weld pool is composed of molten metal coming from the parts being welded, the electrode, and the guide tube. The moving head variation uses an assembly of an electrode guide tube which travels upwards as the weld is laid, keeping it from melting.

Electrogas welding can be applied to most steels, including low and medium carbon steels, low alloy high strength steels, and some stainless steels. Quenched and tempered steels may also be welded by the process, provided that the proper amount of heat is applied. Welds must be vertical, varying to either side by a maximum of 15 degrees. In general, the workpiece must be at least 10 mm (0.4 in) thick, while the maximum thickness for one electrode is approximately 20 mm (0.8 in). Additional electrodes make it possible to weld thicker workpieces. The height of the weld is limited only by the mechanism used to lift the welding head—in general, it ranges from 100 mm (4 in) to 20 m (50 ft).

Like other arc welding processes, EGW requires that the operator wear a welding helmet and proper attire to prevent exposure to molten metal and the bright welding arc. Compared to other processes, a large amount of molten metal is present during welding, and this poses an additional safety and fire hazard. Since the process is often performed at great heights, the work and equipment must be properly secured, and the operator should wear a safety harness to prevent injury in the event of a fall.

Equipment

EGW uses a constant voltage, direct current welding power supply, and the electrode has positive polarity. The welding current can range from 100 to 800 A, and the voltage can range between 30 and 50 V. A wire feeder is used to supply the electrode, which is selected based on the material being welded. The electrode can be flux-cored to provide the weld with protection from atmospheric contamination, or a shielding gas—generally carbon dioxide—can be used with a solid wire electrode. The welding head is attached to an apparatus that elevates during the welding process. Also attached to the apparatus are backing shoes which restrain the weld to the width of the workpieces. To prevent them from melting, they are made of copper and are water-cooled. They must be fit tightly against the joint to prevent leaks.

Diffusion Welding

Animation of the Diffusion Welding process

Diffusion welding (DFW) is a solid state welding process by which two metals (which may be dissimilar) can be bonded together. Diffusion involves the migration of atoms across the joint, due to concentration gradients. The two materials are pressed together at an elevated temperature usually between 50 and 70% of the melting point. The pressure is used to relieve the void that may occur due to the different surface topographies. The method was invented by the Soviet scientist N.F. Kazakov in 1953. Specific tooling is made for each welding application to mate the welder to the workpieces.

Applications

DFW is usually used on sheet metal structures. Typical materials that are welded include titanium, beryllium, and zirconium. It is usually used on low volume workpieces mainly for aerospace, nuclear, and electronics industries.

In many military aircraft diffusion bonding will help to allow for the conservation of expensive strategic materials and the reduction of manufacturing costs. Some aircraft have over 100 diffusion-bonded parts, including; fuselages, outboard and inboard actuator fittings, landing gear trunnions, and nacelle frames.

Atomic Hydrogen Welding

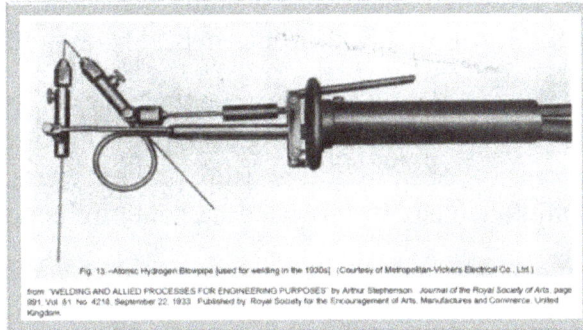

An atomic hydrogen blowpipe, circa 1930.

Atomic hydrogen welding (AHW) is an arc welding process that uses an arc between two metal tungsten electrodes in a shielding atmosphere of hydrogen. The process was invented by Irving Langmuir in the course of his studies of atomic hydrogen. The electric arc efficiently breaks up the hydrogen molecules, which later recombine with tremendous release of heat, reaching temperatures from 3400 to 4000 °C. Without the arc, an oxyhydrogen torch can only reach 2800 °C. This is the third hottest flame after dicyanoacetylene at 4987 °C and cyanogen at 4525 °C. An acetylene torch merely reaches 3300 °C. This device may be called an atomic hydrogen torch, nascent hydrogen torch or Langmuir torch. The process was also known as arc-atom welding.

The heat produced by this torch is sufficient to weld tungsten (3422 °C), the most refractory metal. The presence of hydrogen also acts as a shielding gas, preventing oxidation and contamination by carbon, nitrogen, or oxygen, which can severely damage the properties of many metals. It eliminates the need of flux for this purpose.

The arc is maintained independently of the workpiece or parts being welded. The hydrogen gas is normally diatomic (H_2), but where the temperatures are over 600 °C (1100 °F) near the arc, the hydrogen breaks down into its atomic form, simultaneously absorbing a large amount of heat from the arc. When the hydrogen strikes a relatively cold surface (i.e., the weld zone), it recombines into its diatomic form releasing the energy associated with the formation of that bond. The energy in AHW can be varied easily by changing the distance between the arc stream and the workpiece surface. This process is being replaced by gas metal-arc welding, mainly because of the availability of inexpensive inert gases.

In atomic hydrogen welding, filler metal may or may not be used. In this process, the arc is maintained entirely independent of the work or parts being welded. The work is a part of the electrical circuit only to the extent that a portion of the arc comes in contact with the work, at which time a voltage exists between the work and each electrode.

Laser-hybrid Welding

Laser Hybrid welding is a type of welding process that combines the principles of laser beam welding and arc welding.

The combination of laser light and an electrical arc into an amalgamated welding process has existed since the 1970s, but has only recently been used in industrial applications. There are three main types of hybrid welding process, depending on the arc used: TIG, plasma arc or MIG augmented laser welding. While TIG-augmented laser welding was the first to be researched, MIG is the first to go into industry and is commonly known as hybrid laser welding.

Whereas in the early days laser sources still had to prove their suitability for industrial use, today they are standard equipment in many manufacturing enterprises. The combination of laser welding with another weld process is called a "hybrid welding process". This means that a laser beam and an electrical arc act simultaneously in one welding zone, influencing and supporting each other.

Laser

Laser welding not only requires high laser power but also a high quality beam to obtain the desired "deep-weld effect". The resulting higher quality of beam can be exploited either to obtain a smaller focus diameter or a larger focal distance. A variety of laser types are used for this process, in particular Nd:YAG where the laser light can be transmitted via a water-cooled glass fiber. The beam is projected onto the workpiece by collimating and focusing optics. Carbon dioxide laser can also be used where the beam is transmitted via lens or mirrors.

Laser Hybrid Process

For welding metallic objects, the laser beam is focused to obtain intensities of more than 1 MW/cm^2. When the laser beam hits the surface of the material, this spot is heated up to vaporization temperature, and a vapor cavity is formed in the weld metal due to the escaping metal vapor. This is known as a keyhole. The extraordinary feature of the weld seam is its high depth-to-width ratio. The energy-flow density of the freely burning arc is slightly more than 100 kW/cm^2. Unlike a dual process where two separate weld processes act in succession, hybrid welding may be viewed as a combination of both weld processes acting simultaneously in one and the same process zone. Depending on the kind of arc or laser process used, and depending on the process parameters, the two systems will influence each other in different ways.

The combination of the laser process and the arc process results in an increase in both weld penetration depth and welding speed (as compared to each process alone). The metal vapor escaping from the vapor cavity acts upon the arc plasma. Absorption of the laser radiation in the processing plasma remains negligible. Depending on the ratio of the two power inputs, the character of the overall process may be mainly determined either by the laser or by the arc.

Absorption of the laser radiation is substantially influenced by the temperature of the workpiece surface. Before the laser welding process can start, the initial reflectance must be overcome, especially on aluminum surfaces. This can be achieved by preheating the material. In the hybrid process, the arc heats the metal, helping the laser beam to couple in. After the vaporisation temperature has been reached, the vapor cavity is formed, and nearly all radiation energy can be put into the workpiece. The energy required for this is thus determined by the temperature-dependent absorption and by the amount of energy lost by conduction into the rest of the workpiece. In Laser Hybrid welding, using MIG, vaporisation takes place not only from the surface of the workpiece but also from the filler wire, so that more metal vapor is available to facilitate the absorption of the laser radiation.

Fatigue Behavior

Over the years a great deal of research has been done to understand fatigue behavior, particularly for new techniques like laser hybrid welding, but knowledge is still limited. Laser hybrid welding is an advanced welding technology that creates narrow deep welds and offers greater freedom to control the weld surface geometry. Therefore, fatigue analysis and life prediction of hybrid weld joints has become more important and is the subject of ongoing research.

Cold Welding

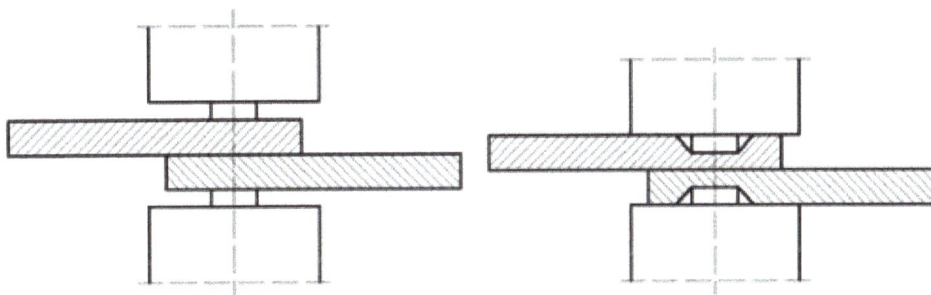

scheme of cold welding

Cold welding or contact welding is a solid-state welding process in which joining takes place without fusion/heating at the interface of the two parts to be welded. Unlike in the fusion-welding processes, no liquid or molten phase is present in the joint.

Cold welding was first recognized as a general materials phenomenon in the 1940s. It was then discovered that two clean, flat surfaces of similar metal would strongly adhere if brought into contact under vacuum. Newly discovered micro- and nano-scale cold welding has already shown great potential in the latest nanofabrication processes.

The reason for this unexpected behavior is that when the atoms in contact are all of the same kind, there is no way for the atoms to "know" that they are in different pieces of copper. When there are other atoms, in the oxides and greases and more complicated thin surface layers of contaminants in between, the atoms "know" when they are not on the same part.

—*Richard Feynman, The Feynman Lectures, 12–2 Friction*

Applications include wire stock and electrical connections (such as insulation-displacement connectors).

In Space

Mechanical problems in early satellites were sometimes attributed to cold welding.

In 2009 the European Space Agency published a peer reviewed paper detailing why cold welding is a significant issue that spacecraft designers need to carefully consider. The paper also cites a documented example from 1991 with the Galileo spacecraft high-gain antenna.

One source of difficulty is that cold welding does not exclude relative motion between the surfaces that are to be joined. This allows the broadly defined notions of galling, fretting, sticking, stiction and adhesion to overlap in some instances. For example, it is possible for a joint to be the result of both cold (or "vacuum") welding and galling (and/or fretting and/or impact). Galling and cold welding, therefore, are not mutually exclusive.

Nanoscale

Unlike cold welding process at macro-scale which normally requires large applied pressures, scientists discovered that single-crystalline ultrathin gold nanowires (diameters less than 10 nm) can be cold-welded together within seconds by mechanical contact alone, and under remarkably low applied pressures. High-resolution transmission electron microscopy and in-situ measurements reveal that the welds are nearly perfect, with the same crystal orientation, strength and electrical conductivity as the rest of the nanowire. The high quality of the welds is attributed to the nanoscale sample dimensions, oriented-attachment mechanisms and mechanically assisted fast surface diffusion. Nanoscale welds were also demonstrated between gold and silver, and silver and silver, indicating that the phenomenon may be generally applicable and therefore offer an atomistic view of the initial stages of macroscopic cold welding for either bulk metals or metallic thin film.

Hyperbaric Welding

Underwater welding

Hyperbaric welding is the process of welding at elevated pressures, normally underwater. Hyperbaric welding can either take place *wet* in the water itself or *dry* inside a specially constructed positive pressure enclosure and hence a dry environment. It is predominantly referred to as "hyperbaric welding" when used in a dry environment, and "underwater welding" when in a wet environment. The applications of hyperbaric welding are diverse—it is often used to repair ships, offshore oil platforms, and pipelines. Steel is the most common material welded.

Dry hyperbaric welding is used in preference to wet underwater welding when high quality welds are required because of the increased control over conditions which can be exerted, such as through application of prior and post weld heat treatments. This improved environmental control leads directly to improved process performance and a generally much higher quality weld than a comparative wet weld. Thus, when a very high quality weld is required, dry hyperbaric welding is normally utilized. Research into using dry hyperbaric welding at depths of up to 1,000 metres (3,300 ft) is ongoing. In general, assuring the integrity of underwater welds can be difficult (but is possible using various nondestructive testing applications), especially for wet underwater welds, because defects are difficult to detect if the defects are beneath the surface of the weld.

Underwater hyperbaric welding was invented by the Russian metallurgist Konstantin Khrenov in 1932.

Application

Welding processes have become increasingly important in almost all manufacturing industries and for structural application. Although a large number of techniques are available for welding in atmosphere, many of these techniques cannot be applied in offshore and marine application where presence of water is of major concern. In this regard, it is relevant to note that a great majority of offshore repairing and surfacing work is carried out at a relatively shallow depth, in the region intermittently covered by the water known as the splash zone. Though numerically, most ship repair and welding jobs are carried out at a shallow depth, the most technologically challenging task is repair at greater depths, especially in pipelines and repair of accidental failure. The advantages of underwater welding are largely of an economic nature, because underwater-welding for marine maintenance and repair jobs bypasses the need to pull the structure out of the sea and saves valuable time and dry docking costs. It is also an important technique for emergency repairs which allow the damaged structure to be safely transported to dry facilities for permanent repair or scrapping. Underwater welding is applied in both inland and offshore environments, though seasonal weather inhibits offshore underwater welding during winter. In either location, surface supplied air is the most common diving method for underwater welders.

Dry Welding

Dry hyperbaric welding involves the weld being performed at raised pressure in a chamber filled with a gas mixture sealed around the structure being welded.

Most arc welding processes such as Shielded Metal Arc Welding (SMAW), Flux-cored arc welding (FCAW), Gas tungsten arc welding (GTAW), Gas metal arc welding (GMAW), Plasma Arc Welding (PAW) could be operated at hyperbaric pressures, but all suffer as the pressure increases. Gas tungsten arc welding is most commonly used. The degradation is associated with physical changes of the arc behaviour as the gas flow regime around the arc changes and the arc roots contract and become more mobile. Of note is a dramatic increase in arc voltage which is associated with the increase in pressure. Overall a degradation in capability and efficiency results as the pressure increases.

Special control techniques have been applied which have allowed welding down to 2,500 m (8,200 ft) simulated water depth in the laboratory, but dry hyperbaric welding has thus far been limited operationally to less than 400 m (1,300 ft) water depth by the physiological capability of divers to operate the welding equipment at high pressures and practical considerations concerning construction of an automated pressure / welding chamber at depth.

Wet Welding

Wet underwater welding directly exposes the diver and electrode to the water and surrounding elements. Divers usually use around 300–400 amps of direct current to power their electrode, and they weld using varied forms of arc welding. This practice commonly uses a variation of shielded metal arc welding, employing a waterproof electrode. Other processes that are used include flux-cored arc welding and friction welding. In each of these cases, the welding power supply is connected to the welding equipment through cables and hoses. The process is generally limited to low carbon equivalent steels, especially at greater depths, because of hydrogen-caused cracking.

Wet welding with a stick electrode is done with similar equipment to that used for dry welding, but the electrode holders are designed for water cooling and are more heavily insulated. They will overheat if used out of the water. A constant current welding machine is used for manual metal arc welding. Direct current is used, and a heavy duty isolation switch is installed in the welding cable at the surface control position, so that the welding current can be disconnected when not in use. The welder instructs the surface operator to make and break the contact as required during the procedure. The contacts should only be closed during actual welding, and opened at other times, particularly when changing electrodes.

The electric arc heats the workpiece and the welding rod, and the molten metal is transferred through the gas bubble around the arc. The gas bubble is partly formed from decomposition of the flux coating on the electrode but it is usually contaminated to some extent by steam. Current flow induces transfer of metal droplets from the electrode to the workpiece and enables positional welding by a skilled operator. Slag deposition on the weld surface helps to slow the rate of cooling, but rapid cooling is one of the biggest problems in producing a quality weld.

Hazards and Risks

The hazards of underwater welding include the risk of electric shock to the welder. To prevent this, the welding equipment must be adaptable to a marine environment, properly insulated and the welding current must be controlled. Commercial divers must also consider the occupational safety issues that divers face; most notably, the risk of decompression sickness due to the increased pressure of breathing gases. Many divers have reported a metallic taste that is related to the galvanic breakdown of dental amalgam. There may also be long term cognitive and possibly musculoskeletal effects associated with underwater welding.

Carbon Arc Welding

Carbon arc welding (CAW) is a process which produces coalescence of metals by heating them with an arc between a nonconsumable carbon (graphite) electrode and the work-piece. It was the first arc-welding process ever developed but is not used for many applications today, having been replaced by twin-carbon-arc welding and other variations. The purpose of arc welding is to form a bond between separate metals. In carbon-arc welding a carbon electrode is used to produce an electric arc between the electrode and the materials being bonded. This arc produces extreme temperatures in excess of 3,000 °C. At this temperature the separate metals form a bond and become welded together.

Development

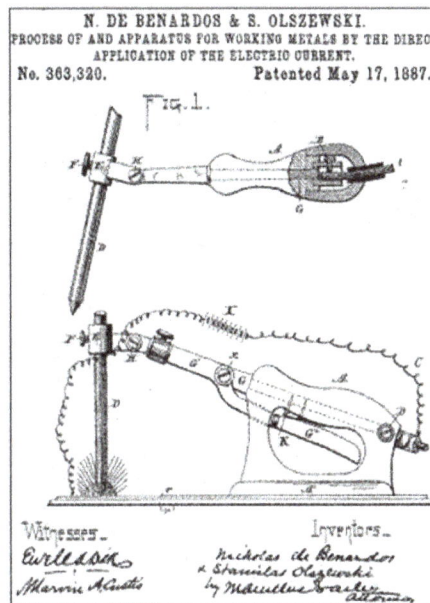

The patent for the arc welding method named *Elektrogefest* ("Electric Hephaestus") granted to Nicholas de Bernardos and Stanisław Olszewski in 1887

CAW could not have been created if not for the discovery of the electric arc by Sir Humphry Davy in 1800, later repeated independently by a Russian physicist Vasily Vladimirovich Petrov in 1802. Petrov studied the electric arc and proposed its possible uses, including welding.

The inventors of carbon-arc welding were Nikolay Benardos and Stanisław Olszewski, who developed this method in 1881 and patented it later under the name *Elektrogefest* ("Electric Hephaestus")

Variations

- Twin carbon arc welding (TCAW) in which the arc is established between two carbon electrodes

- Gas carbon arc welding (CAW-G) no longer has commercial significance. Other processes that use shielding gases have also replaced carbon-arc welding such as *tungsten-arc weld-*

ing (GTAW, or TIG), *plasma-arc welding* (PAW), and *atomic-hydrogen welding* (AHAW). Each of these processes, including carbon-arc welding, use a nonconsumable electrode. A filler metal is generally used to aid the bond in the workpieces.

Exothermic Welding

Thermite welding was a step forward for joining rails

A thermite weld in progress.

Exothermic welding, also known as exothermic bonding, thermite welding (TW), and thermit welding, is a welding that employs molten metal to permanently join the conductors. The process employs an exothermic reaction of a thermite composition to heat the metal, and requires no external source of heat or current. The chemical reaction that produces the heat is an aluminothermic reaction between aluminium powder and a metal oxide.

Overview

In exothermic welding, aluminium dust reduces the oxide of another metal, most commonly iron oxide, because aluminium is highly reactive. Iron(III) oxide is commonly used:

$$Fe_2O_3 + 2\ Al2\ Fe + Al_2O_3$$

The products are aluminium oxide, free elemental iron, and a large amount of heat. The reactants are commonly powdered and mixed with a binder to keep the material solid and prevent separation.

Commonly the reacting composition is five parts iron oxide red (rust) powder and three parts aluminium powder by weight, ignited at high temperatures. A strongly exothermic (heat-generating) reaction occurs that via reduction and oxidation produces a white hot mass of molten iron and a slag of refractory aluminium oxide. The molten iron is the actual welding material; the aluminium oxide is much less dense than the liquid iron and so floats to the top of the reaction, so the set-up for welding must take into account that the actual molten metal is at the bottom of the crucible and covered by floating slag.

Other metal oxides can be used, such as chromium oxide, to generate the given metal in its elemental form. Copper thermite, using copper oxide, is used for creating electric joints:

$$3 Cu_2O+2Al6 Cu+Al_2O_3$$

Thermite welding is widely used to weld railway rails. One of the first railroads to evaluate the use of thermite welding was the Delaware Hudson in 1935 The weld quality of chemically pure thermite is low due to the low heat penetration into the joining metals and the very low carbon and alloy content in the nearly pure molten iron. To obtain sound railroad welds, the ends of the rails being thermite welded are preheated with a torch to an orange heat, to ensure the molten steel is not chilled during the pour. Because the thermite reaction yields relatively pure iron, not the much stronger steel, some small pellets or rods of high-carbon alloying metal are included in the thermite mix; these alloying materials melt from the heat of the thermite reaction and mix into the weld metal. The alloying beads composition will vary, according to the rail alloy being welded.

The reaction reaches very high temperatures, depending on the metal oxide used. The reactants are usually supplied in the form of powders, with the reaction triggered using a spark from a flint lighter. The activation energy for this reaction is very high however, and initiation requires either the use of a "booster" material such as powdered magnesium metal or a very hot flame source. The aluminium oxide slag that it produces is discarded.

When welding copper conductors, the process employs a semi-permanent graphite crucible mould, in which the molten copper, produced by the reaction, flows through the mould and over and around the conductors to be welded, forming an electrically conductive weld between them. When the copper cools, the mould is either broken off or left in place. Alternatively, hand-held graphite crucibles can be used. The advantages of these crucibles include portability, lower cost (because they can be reused), and flexibility, especially in field applications.

Properties

An exothermic weld has higher mechanical strength than other forms of weld, and excellent corrosion resistance It is also highly stable when subject to repeated short-circuit pulses, and does not suffer from increased electrical resistance over the lifetime of the installation. However, the process is costly relative to other welding processes, requires a supply of replaceable moulds, suffers from a lack of repeatability, and can be impeded by wet conditions or bad weather (when performed outdoors).

Applications

Exothermic welding is usually used for welding copper conductors but is suitable for welding a wide range of metals, including stainless steel, cast iron, common steel, brass, bronze, and Monel. It is especially useful for joining dissimilar metals. The process is marketed under a variety of names such as APLIWELD (in tablet form), American Rail Weld, Harger ULTRASHOT, ERICO CADWELD, Quikweld, Tectoweld, Ultraweld, Techweld, TerraWeld, Thermoweld, Ardo Weld, AmiableWeld, AIWeld, FurseWeld and Kumwell.

Because of the good electrical conductivity and high stability in the face of short-circuit pulses, exothermic welds are one of the options specified by §250.7 of the United States National Electrical Code for grounding conductors and bonding jumpers. It is the preferred method of bonding, and indeed it is the only acceptable means of bonding copper to galvanized cable. The NEC does not require such exothermically welded connections to be listed or labelled, but some engineering specifications require that completed exothermic welds be examined using X-ray equipment.

Rail Welding

Tram tracks being joined

Tram tracks recently joined

Typically, the ends of the rails are cleaned, aligned flat and true, and spaced apart 25 mm (1 in). This gap between rail ends for welding is to ensure consistent results in the pouring of the molten steel into the weld mold. In the event of a welding failure, the rail ends can be cropped to a 75 mm (3 in) gap, removing the melted and damaged rail ends, and a new weld attempted with a special mould and larger thermite charge. A two or three piece hardened sand mould is clamped around the rail ends, and a torch of suitable heat capacity is used to preheat the ends of the rail and the interior of the mould. The proper amount of thermite with alloying metal is placed in a refractory crucible, and when the rails have reached a sufficient temperature, the thermite is ignited and allowed to react to completion (allowing time for any alloying metal to fully melt and mix, yielding the desired molten steel or alloy). The reaction crucible is then tapped at the bottom. Modern crucibles have a self-tapping thimble in the pouring nozzle. The molten steel flows into the mould, fusing with the rail ends and forming the weld. The slag, being lighter than the steel flows last from the crucible and overflows the mould into a steel catch basin, to be disposed of after cooling. The entire setup is allowed to cool. The mould is removed and the weld is cleaned by hot chiselling and grinding to produce a smooth joint. Typical time from start of the work until a train can run over the rail is approximately 45 minutes to more than an hour, depending on the rail size and ambient temperature. In any case, the rail steel must be cooled to less than 370 °C (700 °F) before it can sustain the weight of rail locomotives.

When a thermite process is used for track circuits – the bonding of wires to the rails with a copper alloy, a graphite mould is used. The graphite mould is reusable many times, because the copper alloy is not as hot as the steel alloys used in rail welding. In signal bonding, the volume of molten copper is quite small, approximately 2 cm^3 (0.1 cu in) and the mould is lightly clamped to the side of the rail, also holding a signal wire in place. In rail welding, the weld charge can weigh up to 13 kg (29 lb). The hardened sand mould is heavy and bulky, must be securely clamped in a very specific position and then subjected to intense heat for several minutes before firing the charge. When rail is welded into long strings, the longitudinal expansion and contraction of steel must be taken into account. British practice sometimes uses a sliding joint of some sort at the end of long runs of continuously welded rail, to allow some movement, although by using a heavy concrete sleeper and an extra amount of ballast at the sleeper ends, the track, which will be prestressed according to the ambient temperature at the time of its installation, will develop compressive stress in hot ambient temperature, or tensile stress in cold ambient temperature, its strong attachment to the heavy sleepers preventing buckling or other deformation. Current practice is to use welded rails throughout on high speed lines, and expansion joints are kept to a minimum, often only to protect junctions and crossings from excessive stress. American practice appears to be very similar, a straightforward physical restraint of the rail. The rail is prestressed, or considered "stress neutral" at some particular ambient temperature. This "neutral" temperature will vary according to local climate conditions, taking into account lowest winter and warmest summer temperatures. The rail is physically secured to the ties or sleepers with rail anchors, or anti-creepers. If the track ballast is good and clean and the ties are in good condition, and the track geometry is good, then the welded rail will withstand ambient temperature swings normal to the region.

Remote Welding

Remote exothermic welding is a type of exothermic welding process for joining two electrical conductors from a distance. The process reduces the inherent risks associated with exothermic welding and is used in installations that require a welding operator to permanently join conductors from a safe distance of the superheated copper alloy.

The process incorporates either an igniter for use with standard graphite molds or a consumable sealed drop in weld metal cartridge, semi-permanent graphite crucible mold, and an ignition source that tethers to the cartridge with a cable that provides the safe remote ignition.

X-ray Welding

X-ray Welding is an experimental welding process that uses a high powered X-ray source to provide thermal energy required to weld materials.

The phrase "X-ray welding" also has an older, unrelated usage in quality control. In this context, an X-Ray welder is a tradesman who consistently welds at such a high proficiency that he rarely introduces defects into the weld pool, and is able to recognize and correct defects in the weld pool, during the welding process. It is assumed (*or trusted*) by the Quality Control Department of a fabrication or manufacturing shop that the welding work performed by an X-ray welder would pass an X-ray inspection. For example, defects like porosity, concavities, cracks, cold laps, slag and tungsten inclusions, lack of fusion & penetration, etc., are rarely seen in a radiographic X-ray inspection of a weldment performed by an X-ray welder.

With the growing use of synchrotron radiation in the welding process, the older usage of the phrase "X-Ray welding" might cause confusion; but the two terms are unlikely to be used in the same work environment because synchrotron radiation (X-Ray) welding is a remotely automated and mechanized process.

Introduction

Many advances in welding technology have resulted from the introduction of new sources of the thermal energy required for localized melting. These advances include the introduction of modern techniques such as gas tungsten arc, gas-metal arc, submerged-arc, electron beam, and laser beam welding processes. However, whilst these processes were able to improve stability, reproducibility, and accuracy of welding, they share a common limitation - the energy does not fully penetrate the material to be welded, resulting in the formation of a melt pool on the surface of the material.

To achieve welds which penetrate the full depth of the material, it is necessary to either specially design and prepare the geometry of the joint or cause vaporization of the material to such a degree that a "keyhole" is formed, allowing the heat to penetrate the joint. This is not a significant disadvantage in many types of material, as good joint strengths can be achieved, however for certain material classes such as ceramics or metal ceramic composites, such processing can significantly limit joint strength. They have great potential for use in the aerospace industry, provided a joining process that maintains the strength of the material can be found.

Until recently, sources of x-rays of sufficient intensity to cause enough volumetric heating for welding were not available. However, with the advent of third-generation synchrotron radiation sources, it is possible to achieve the power required for localized melting and even vaporization in a number of materials.

X-ray beams have been shown to have potential as welding sources for classes of materials which cannot be welded conventionally.

Laser Guided and Stabalized Arc Welding

Laser guided and stabilized welding (LGS-welding) is a process in which a laser beam irradiates an electrical heated plasma arc to set a path of increased conductivity. Therefore, the arc´s energy can be spatial directed and the plasma burns more stable. The process must be destincted from laser-hybrid welding, since only low power laser energy of a couple hundred watts is used and the laser does not contribute significantly to the welding process in terms of energy input.

Operation

The principle of Laser enhanced welding is based on the interaction between the electrical arc and laser radiation. Due to the Optogalvanic effect (OGE) a channel of higher conductivity in the plasma is established along the path of the laser. Therefore, a movement of the laser beam results in a movement of the electrical arc. This effect is limited to a range of some milimeters, but shows the influence of the radiation to the plasma. A raise of welding speed of over 100% is described by using a diode laser with a wavelength of 811 nm without a significant loose in penetration depth. Furthermore, this technique is used in cladding. Depending on the welded material argon or argon with CO_2 is used as shielding gas. The laser source must be tuned to emit at a wavelength of 811 nm and is focused into the plasma.

Laser Guided and Stabilized GMA-Welding

The process is used for welding thin metal sheets up to about 2 mm when welding in overlap or butt joint. LGS-GMA-welding is most advantageous when welding fillet welds. The guidance effect of the laser radiation forces the arc into the fillet. Therefore, a steady seam can be reached. Furthermore, the stabilization of the plasma enables the GMA-process to weld thin sheets without burning holes in the material.

Equipment and Setup

The setup requires the GMA welding head tilted at 60° to the work piece surface. In order to realize a maximum overlap between the electric arc and the laser beam in the process area, the laser is installed upright to the workpiece and focused in the electrical arc. Standard welding equipment can be used for the process. The laser source is described above.

Laser Guided and Stabilized Double Head TIG-welding

In laser guided and stabilized double head TIG-welding the laser forces two arcs together. The goal of this technique is to increase welding speed of TIG-welding without compromising the quality.

Equipment and Setup

For this process two TIG-sources are needed and the laser described above. The TIG-torches are set up with the laser beam perpendicular in the middle. All welding modes of the two torches are possible (DC/DC,AC/AC,AC/DC).

Laser Guided and Stabiliszed GMA-Cladding

In LGS-GMA-cladding the stabilization effect is used enable the GMA-process to work with low energy. This is needed to reduce the penetration depth an therefore the dilution of base and deposition material. The combination of GMA-welding and a diode laser leads to a cheap and energy efficient process.

Equipment and Setup

The setup for the LGS-GMA-cladding is almost alike the one for LGS-GMA-welding beside that the GMA-source needs to have a "Cold-MIG" process. This means, that the welding current is controlled my microcontrollers and produced by power electronics. That way not only the current peaks can be controlled, but also the slopes.

References

- Sacks, Raymond; Bohnart, E. (2005). "17". Welding Principles and Practices (Third ed.). New York: McGraw_Hill. p. 597. ISBN 978-0-07-825060-6.

- Kalpakjian, Serope; Schmid, Steven R (2006). Manufacturing Engineering and Technology (5th ed.). Prentice Hall. ISBN 978-0-13-187599-9.

- Cary, HB; Helzer, SC (2005). Modern Welding Technology. Upper Saddle River, New Jersey: Pearson Education. pp. 677–681. ISBN 0-13-113029-3.

- Carl W. Hall A biographical dictionary of people in engineering: from the earliest records until 2000, Vol. 1, Purdue University Press, 2008 ISBN 1-55753-459-4 p. 120

- Bevan, John, ed. (2005). "Section 3.3". The Professional Divers's Handbook (second ed.). 5 Nepean Close, Alverstoke, GOSPORT, Hampshire PO12 2BH: Submex Ltd. pp. 122–125. ISBN 978-0950824260.

- Messler, Robert (2004). Joining of materials and structures : from pragmatic process to enabling technology. Elsevier. p. 296. ISBN 978-0-7506-7757-8.

- Milenko Braunović; Valeriĭ Vasil'evich Konchits; Nikolaĭ Konstantinovich Myshkin (2006). Electrical Contacts: Fundamentals, Applications and Technology. CRC Press. p. 291. ISBN 9781574447279.

Diverse Techniques of Welding

A welding power supply is a device and this device helps in performing welding. It is a very simple device, it is as simple as a car battery. Along with welding power supply, shot welding, welding procedure specification, weld purging and orbital welding are the essential techniques of welding. The section strategically encompasses and incorporates the major techniques related to welding, providing a complete understanding.

Welding Power Supply

A welding power supply is a device that provides an electric current to perform welding. Welding usually requires high current (over 80 amperes) and it can need above 12,000 amperes in spot welding. Low current can also be used; welding two razor blades together at 5 amps with gas tungsten arc welding is a good example. A welding power supply can be as simple as a car battery and as sophisticated as a high-frequency inverter using IGBT technology, with computer control to assist in the welding process.

Classification

Welding machines are usually classified as constant current (CC) or constant voltage (CV); a constant current machine varies its output voltage to maintain a steady current while a constant voltage machine will fluctuate its output current to maintain a set voltage. Shielded metal arc welding and gas tungsten arc welding will use a constant current source and gas metal arc welding and flux-cored arc welding typically use constant voltage sources but constant current is also possible with a voltage sensing wire feeder.

The nature of the CV machine is required by gas metal arc welding and flux-cored arc welding because the welder is not able to control the arc length manually. If a welder attempted to use a CV machine to weld with shielded metal arc welding the small fluctuations in the arc distance would cause wide fluctuations in the machine's output. With a CC machine the welder can count on a fixed number of amps reaching the material to be welded regardless of the arc distance but too much distance will cause poor welding.

Power Supply Designs

The welding power supplies most commonly seen can be categorized within the following types:

Transformer

A transformer-style welding power supply converts the moderate voltage and moderate current electricity from the utility mains (typically 230 or 115 VAC) into a high current and low voltage supply, typically between 17 and 45 (open-circuit) volts and 55 to 590 amperes. A rectifier converts the AC into DC on more expensive machines.

This design typically allows the welder to select the output current by variously moving a primary winding closer or farther from a secondary winding, moving a magnetic shunt in and out of the core of the transformer, using a series saturating reactor with a variable saturating technique in series with the secondary current output, or by simply permitting the welder to select the output voltage from a set of taps on the transformer's secondary winding. These transformer style machines are typically the least expensive.

The trade off for the reduced expense is that pure transformer designs are often bulky and massive because they operate at the utility mains frequency of 50 or 60 Hz. Such low frequency transformers must have a high magnetizing inductance to avoid wasteful shunt currents. The transformer may also have significant leakage inductance for short circuit protection in the event of a welding rod becoming stuck to the workpiece. The leakage inductance may be variable so the operator can set the output current.

Generator and Alternator

Welding power supplies may also use generators or alternators to convert mechanical energy into electrical energy. Modern designs are usually driven by an internal combustion engine but older machines may use an electric motor to drive an alternator or generator. In this configuration the utility power is converted first into mechanical energy then back into electrical energy to achieve the step-down effect similar to a transformer. Because the output of the generator can be direct current, or even a higher frequency AC, these older machines can produce DC from AC without any need for rectifiers of any type, or can also be used for implementing formerly-used variations on so-called heliarc (most often now called TIG) welders, where the need for a higher frequency add-on module box is avoided by the alternator simply producing higher frequency ac current directly.

Inverter

Since the advent of high-power semiconductors such as the insulated gate bipolar transistor (IGBT), it is now possible to build a switched-mode power supply capable of coping with the high loads of arc welding. These designs are known as inverter welding units. They generally first rectify the utility AC power to DC; then they switch (invert) the DC power into a stepdown transformer to produce the desired welding voltage or current. The switching frequency is typically 10 kHz or higher. Although the high switching frequency requires sophisticated components and circuits, it drastically reduces the bulk of the step down transformer, as the mass of magnetic components (transformers and inductors) that is required for achieving a given power level goes down rapidly as the operating (switching) frequency is increased. The inverter circuitry can also provide features such as power control and overload protection. The high frequency inverter-based welding machines are typically more efficient and provide better control of variable functional parameters than non-inverter welding machines.

The IGBTs in an inverter based machine are controlled by a microcontroller, so the electrical characteristics of the welding power can be changed by software in real time, even on a cycle by cycle basis, rather than making changes slowly over hundreds if not thousands of cycles. Typically, the controller software will implement features such as pulsing the welding current, providing variable ratios and current densities through a welding cycle, enabling swept or stepped variable frequencies, and providing timing as needed for implementing automatic spot-welding; all of these features would be prohibitively expensive to design into a transformer-based machine, but require only program memory space in a software-controlled inverter machine. Similarly, it is possible to add new features to a software-controlled inverter machine if needed, through a software update, rather than through having to buy a more modern welder.

Other Types

Additional types of welders also exist, besides the types using transformers, motor/generator, and inverters. For example, laser welders also exist, and they require an entirely different type of welding power supply design that does not fall into any of the types of welding power supplies discussed previously. Likewise, spot welders require a different type of welding power supply, typically containing elaborate timing circuits and large capacitor banks that are not commonly found with any other types of welding power supplies.

Shot Welding

Stainess steel card with two examples of shot welding given as souvenirs from the Budd Company circa 1934.

Shot welding is a specific type of spot welding used to join two pieces of metal together. This is accomplished by clamping the two pieces together and then passing a large electric current through them for a short period of time. Assuming the right amount of current for the right time, this will weld the two pieces of metal together. Shot welding was invented by Earl J. Ragsdale, a mechanical engineer at the Budd Company, in 1932 for the purposes of welding stainless steel. This welding method was used to construct the *Pioneer Zephyr*.

The Method

The E. G. Budd Company of Philadelphia recognized the important metallurgical characteristics of 18/8 stainless steel (known today as SAE 304 austenitic stainless steel) and developed a spot welding process to take advantage of the oxidized layer on the surface of stainless steel. Heat treating the 18-8 stainless steel leaves a metal with non-magnetic and ductile properties. Repeatedly reheating the metal to 1000–1100 °C impairs the mechanical and chemical properties of the metal. The metal becomes susceptible to corrosion due to carbide precipitation, and loses fatigue resistance. The important factor in controlling the metal's properties is the dwell time at those temperatures. Using a controlled time element and recorder, a power supply with smooth current, and very brief high currents, a satisfactory spot weld may be produced.

The corona of the shot weld should not exist on the metal, and the equipment used produces satisfactory welds with a smaller than normal diameter. Sufficient electrode force is applied to hold the two sheets of metal together and the peak current rapidly creates a forge weld at the interface between the two sheets, producing a small nugget of weld metal, which when cooled results in a shear resistant metal interface. Good shotwelds have twice the shear strength of a rivet of similar diameter and can be placed 50 percent closer together. Distortion is eliminated, as this is an issue in fusion welding processes.

Welding Procedure Specification

Bend test coupons for welding procedure qualification.

A Welding Procedure Specification (WPS) is the formal written document describing welding procedures, which provides direction to the welder or welding operators for making sound and quality production welds as per the code requirements . The purpose of the document is to guide welders to the accepted procedures so that repeatable and trusted welding techniques are used. A WPS is developed for each material alloy and for each welding type used. Specific codes and/or engineering societies are often the driving force behind the development of a company's WPS. A WPS is supported by a Procedure Qualification Record (PQR or WPQR). A PQR is a record of a test weld performed and tested (more rigorously) to ensure that the procedure will produce a good weld. Individual welders are certified with a qualification test documented in a Welder Qualification Test Record (WQTR) that shows they have the understanding and demonstrated ability to work within the specified WPS.

Introduction

The following are definitions for WPS and PQR found in various codes and standards:

According to the American Welding Society (AWS), a WPS provides in detail the required welding variables for specific application to assure repeatability by properly trained welders. The AWS defines welding PQR as a record of welding variables used to produce an acceptable test weldment and the results of tests conducted on the weldment to qualify a Welding Procedure Specification. For steel construction (civil engineering structures) AWS D1.1 is a widely used standard. It specifies either a *pre-qualification* option (chapter 3) or a *qualification* option (chapter 4) for approval of welding processes.

The American Society of Mechanical Engineers (ASME) similarly defines a WPS as a written document that provides direction to the welder or welding operator for making production welds in accordance with Code requirements. ASME also defines welding PQR as a record of variables recorded during the welding of the test coupon. The record also contains the test results of the tested specimens.

The Canadian Welding Bureau, through CSA Standards W47.1, W47.2 and W186, specifies both a WPS and a Welding Procedure Data Sheet (WPDS) to provide direction to the welding supervisor, welders and welding operators. The WPS provides general information on the welding process and material grouping being welded, while the WPDS provides specific welding variables/parameter/conditions for the specific weldment. All WPS and WPDS must be independently reviewed and accepted by the Canadian Welding Bureau prior to use. These CSA standards also define requirements for procedure qualification testing (PQT) to support the acceptance of the WPDS. A record of the procedure qualification test and the results must be documented on a procedure qualification record (PQR). All PQTs are independently witnessed by the Canadian Welding Bureau.

In Europe, the European Committee for Standardization (CEN) has adopted the ISO standards on welding procedure qualification (ISO 15607 to ISO 15614), which replaced the former European standard EN 288. EN ISO 15607 defines a WPS as "A document that has been qualified by one of the methods described in clause 6 and provides the required variables of the welding procedure to ensure repeatability during production welding". The same standard defines a *Welding Procedure Qualification Record* (WPQR) as "Record comprising all necessary data needed for qualification of a preliminary welding procedure specification". In addition to the standard WPS qualification procedure specified in ISO 15614, the ISO 156xx series of standards provides also for alternative WPS approval methods. These include: *Tested welding consumables* (ISO 15610), *Previous welding experience* (ISO 15611), *Standard welding procedure* (ISO 15612) and *Preproduction welding test* (ISO 15613).

In the oil and gas pipeline sector, the American Petroleum Institute API 1104 standard is used almost exclusively worldwide. API 1104 accepts the definitions of the American Welding Society code AWS A3.0.

Weld Purging

Weld purging. is the act of removing, from the vicinity of the joint, oxygen, water vapour and any other gases or vapours that might be harmful to a welding joint as it is being welded and immediately after welding.

Stainless steels, duplex steels, titanium-, nickel- and zirconium- alloys are sensitive to the presence of air, oxygen, hydrogen, water vapour and other vapours and gases that may combine with the hot metal as it is being joined.

Such gases may combine with the metal to form undesirable compounds that may reduce corrosion resistance or may be instrumental in creating cracks or other structural defects in metals.

Weld Purging is generally necessary for the first weld run when joining to separate parts. This sealing weld will be called a "root run" when it takes more than one run or (pass) to fully seal the root area from above.

Once the root run has been completed, it is possible to stop the purging process unless the welding engineer has specified that purging should be continued for the second and third passes for example in case the root weld becomes hot enough to oxidise in the air that will have replaced the purge gas.

Purging Methods

Unwanted gas is generally removed by flushing with an inert gas. Argon is generally used for this purpose but helium is an alternative depending on gas cost and availability.

Nitrogen has been used as a purge gas but is unsuitable for some stainless steels.[

The most common way to remove gas from the weld zone is to flush it away with an inert gas. The weld zone can be contained to prevent fresh gas from entering once the contained volume has been purged.

Another method of purging is to enclose the metal parts completely in a vacuum chamber and evacuate it, prior to backfilling with inert gas for the welding process. When purging with inert gases, it is important to enter the gas very slowly.

With Argon, which is heavier than air, it should be entered from the bottom of the enclosed space, where it will move slowly across the base area, be it the bottom of a pipe or tank, and then move slowly upwards displacing the air upwards out of a release hole at the top. This is where a Weld Purge Monitor®, oxygen sensing instrument, should be placed.

Helium however, which is also an inert gas should be entered from the top of the enclosed space, where it will first flood across the upper limits and push the trapped air downwards and out through an exit where the oxygen level can be measured until it reaches the required level. The purpose of slow gas entry, typically 5 – 7 litres/m (....CFH), is to avoid turbulence, which would otherwise lengthen the purge process considerably.

Welding Methods

Most weld purging is carried out on joints made by the TIG or GTAW welding process. When the laser welding process is chosen, joints or welds being made on reactive materials will need to be carried out and in the case of electron beam welding, this is carried out in a vacuum, in which case purging takes place by complete evacuation of all gas.

For TIG/GTAW welding, the top side (front side) of a weld is normally protected by inert gas flowing through the welding torch and it is the underside (back side) of the weld where atmospheric gases need to be purged.

When joining two parts together with a seal weld, it may be necessary to first hold them together, by making tack welds. This will be carried out by making very short welds at intervals varying according to the size and weight of the two parts. It is very important that the weld purging process should start for the tack welds, so the undersides of those tacks remain clean and shiny, without any oxidation or discolouration.

Gases

Unwanted gas is generally removed by flushing with an inert gas. Argon is generally used for this purpose but helium is an alternative depending on gas cost and availability. Nitrogen has been used as a purge gas but is unsuitable for some stainless steels. Purge gases must be of a certain quality, in order that the welds are made correctly. Because welds in Argon or Helium which have more than 0.05% (500 ppm) of oxygen will oxidise and discolour, welding engineers will include the mention of the correct gas quality in their welding procedures.

It is important to use a Weld Purge Monitor oxygen measuring instrument as a means to inspect the purge gas before purging is actually started.

Measurement of Purge Quality

There are charts available that show the discolouration of some metals, caused by the presence of oxygen, even at concentrations below 50 parts per million. It is possible to monitor oxygen levels by an instrument known as a Weld Purge Monitor. There are different monitors available to measure oxygen levels accurately between ambient levels and 1 part per million which is low enough to prevent contamination of the most sensitive materials.

Some of these instruments are hand held and powered by battery for ease of use around construction sites where the materials mentioned are being welded and others are powered by mains electricity due to the power requirements needed to run switching operations, collect data and operate internal pumps to draw the purge gas over the sensor for measurement.

Accessories

To carry out weld purging, there are many proprietary accessories available to suit all kinds of weld joints.

Tube and Pipe Welds With Open Ended Access

Inflatable pipe purging systems are the most commonly used, where two inflatable bags (known as dams) are connected by an inert gas tube about 20" (500 mm) in length with one dam placed either side of the weld.

The space between the dams is filled with inert gas that flushes out the air. The residual oxygen is measured by a weld purge monitor until it reaches the desired level for welding to begin.

Closing Tube and Pipe Welds

In some cases it is not possible to retrieve reusable pipe purging systems so a water-soluble film is the best alternative. Film dams can be placed either side of a weld and held in place with a strong water-soluble adhesive. Water-soluble dams are transparent, allowing welders and quality control engineers to view and check the weld from the underside while it is in progress. After welding, the Dams are simply washed away during water cleaning or hydrostatic leak testing of joints.

Sheet Metal Joints

Sheet metal joints might be made on a seam welding machine where weld purging is carried out automatically as part of the machine design.

Where sheet metal is formed into tanks or vessels prior to welding, the inside of the vessel can be purged with a flow of inert gas, however for larger sizes the cost of the gas and the time taken is unrealistic. In such cases, a weld backing tape can be used. This is a layer of glass fibre band in the centre of a width of adhesive aluminium foil that is placed over the rear of the weld joint.

Welds made onto weld backing tape can be carried out faster than normal and the weld bead is cast flat onto the glass fibre leaving an acceptable weld profile behind. This method may leave some discolouration to be cleaned, but dramatically reduces the amount of coking and oxidation that would otherwise take place.

Trailing Shields

To improve weld purging on the front side of welds, a trailing shield can be attached to manual or automatic welding torches to follow behind as the weld is being made. In this way, the weld is kept under an inert gas shield for longer than normal, giving an improved weld quality, while allowing the welder to weld faster.

Trailing shields are also supplied to hold against the back side of welds, whether flat or radiused, in order to keep the joint free from oxidation and discolouration.

When welding with metals such as Titanium, a secondary inert shielding gas application is necessary to protect the cooling weld bead and heat-affected zone. Trailing cups prevent oxidation by shielding the weld from the atmosphere until it has cooled to a safe temperature.

Chambers

When separate components need to be welded in an inert atmosphere, they can be placed inside a Weld Purging Chamber that is flushed out completely with inert gas.

Welding chambers are mostly built to order, however standard low cost Flexible Welding Enclosures are available, that enable individual or multiples of individual components to be welded in one weld purge cycle.

Technicalities

Argon is heavier than air. When used as a purge gas, it should be slowly introduced into the bottom of the cavity and allowed to push the ambient gas out of the exhaust port. This method avoids the ambient gas and the purge gas mixing and which would lengthen the weld purge cycle considerably.

Helium is lighter than air and so has to be introduced at the top of the cavity with the exhaust at the bottom. For better results a weld purge monitor should be used rather than guessing whether the purge is satisfactory. One weld that has to be cut out and repaired, will always cost more than such an instrument.

Cleanliness

To ensure an easy purge cycle and a good weld being made, cleanliness has to be considered as part of the weld preparation.

All surfaces adjacent to the weld and in a trapped space that is being purged, should be thoroughly cleaned with a permitted industrial cleaning fluid, dried and washed afterwards in demineralized water.

Welders and pipe fitters should try to ensure that the purge cavity is always free from oils, paper, moisture, cloth, foam, sponge etc. All such items will give off vapours continually during the weld cycle and could risk contamination of the weld joint and rejection or failure of the weld when in service.

Weld purging products suitable for this work should use specially selected materials for construction with lowest possible vapour pressures to avoid outgassing when the weld zone starts to become warm from welding.

As the weld warms the immediate vicinity inside the weld purge space, any materials exposed to the increasing temperature will start to outgas (give off gas).

Normally this extra gas should be removed by the continuing weld purge gas flow, however there are times when gas and vapours overwhelm the purge, oxygen levels rise and the weld becomes contaminated.

This should be spotted by the welder with his weld purging instrument.

Some advanced Weld Purge Monitor oxygen sensing instruments have the ability to give an alarm signal or close down the welding operation, when there is an unwanted rise in oxygen level.

Purging can continue until the oxygen level is low enough once more and welding can begin again.

To avoid such situations, undesirable materials must be kept away from the weld zone.

Orbital Welding

Orbital welding is a specialized area of welding whereby the arc is rotated mechanically through 360° (180 degrees in double up welding) around a static workpiece, an object such as a pipe, in a continuous process. The process was developed to addresses the issue of operator error in gas tungsten arc welding processes (GTAW). In orbital welding, computer-controlled process runs with little intervention from the operator. The process is used specifically for high quality repeatable welding.

History

The orbital welding process was invented by Chief of Research, V.H. Pavlecka, and engineer Russ Meredith of Northrup Aircraft Inc. over 50 years ago to address the issue of operator error in GTAW.

Equipment

The main components of every orbital welding system are the power source and controller, the welding head and, where required, a wire feed mechanism. Welding of certain sizes and material types will also require the use of a water/coolant system. There are a large number of factors that can have an influence on the welding result. These aspects include the arc length, magnitude and pulse frequency of the welding current, welding speed, inert shielding gas, parent material, filler material, weld preparation, and thermal conductivity. Ultimately, a high quality weld is achieved through detailed knowledge of how to precisely adjust all these parameters for each individual welding task.

Application

The Welding Process

It is very difficult to achieve the highest standards of quality and safety using manual welding. This is due to certain welding positions, overhead and down-hand welds for example, often leading to faulty welds due to restricted access the user has in these welding positions. In order to have complete control over the weld pool, a perfect balance must be maintained between gravitational force and surface tension at every position of the torch. By using mechanised variants of the technique, certain parts of the welding process are handled by mechanical components. Note that a welder will always be monitoring and controlling the process. In an ideal situation, all welding parameters would be fully programmed before welding is started. In practice, however, the presence of variable constraints means that it is often necessary for the welder to make corrective interventions.

Orbital Welding of Tubing using totally enclosed weld heads is a fusion process under ASME Section 9. No filler metal is added.

A successful automatic orbital GTA weld is 100% repeatable as long as the operator monitors variables and performs periodic samples or coupons which are inspected for complete penetration. Noticing that a variable has changed is a primary skill and can be easily missed. Training and experience are required for an operator to be successful at consistently producing acceptable welds.

ASME requirements for certification of a person as an orbital welding operator requires the person to set up the weld head and program the welding machine and produce 6 consecutive samples that will pass bend and/or tensile testing. Professional instruction is typically obtained for a person to be able to make samples that pass the test.

The successful automatic orbital GTA weld is very dependent upon refinement of several critical variables that involve programming the welding machine and set-up of the "weld head".

Maintenance of the weld head often becomes a factor in repeatability of successful welds. Weld head internals can become charred from improper use. The charring is carbon deposits which can conduct electricity and short circuit the flow of current from the tungsten. Weld heads contain a system of precision planetary gears that can wear out over time. Proper cleaning and maintenance is required.

Successful orbital welding is also dependent upon using high quality tubing material. Typically only 316L stainless steel tubing (not pipe) and fittings are used for automatic orbital GTA welding and are obtained from a number of specialty manufacturers.

Successful orbital welding is also dependent upon having a reasonably clean source of Argon for backing and shielding gas. Minimum purity would be 99.995% for typical industrial applications. For some applications it is necessary to use ultra high purity argon, 99.9998% purity and such applications requires the use of all high purity purge equipment (valves, regulators and flow control). Typically, no rubber components can be used for purge gas apparatus since the rubber absorbs and releases moisture and oxygen into the argon stream. Moisture and oxygen (in Argon) are contaminants detrimental to a successful automatic orbital weld.

Weld coupons, pieces of metal used to test a welders' skill, are typically prepared at the beginning of a welding shift, any time any variable is adjusted or changed and at the end of the shift (and more frequently as required by an inspector). Each coupon must be examined internally and externally to verify full penetration, proper bead width and other criteria. With smaller diameters it is usually necessary to section open the coupon to examine the weld bead. All coupons must exhibit complete penetration and consistent bead width. Variations in consistency are an indicator of a problem that must be resolved before continuing.

Orbital welding is typically only performed on TUBING and not pipe for several reasons most important being that the production of tubing yields very consistent outside diameters which is critical to proper fit up in the weld head.

Automatic Orbital GTA welding has become the standard joining method for high integrity gas and liquid systems used in the SEMICONDUCTOR and Pharmaceutical manufacturing industries. These systems are rated for extreme purity and leak tight integrity. An entire specialty industry suppling valves, fittings, regulators, gauges and other components for orbital welding and use in high purity applications has developed since the mid 1980s. For tube welding in high purity applications only a fully enclosed weld head may be used.

Materials

Orbital welding has almost always exclusively been carried out by the Tungsten Inert Gas (TIG / GTAW) technique using non-consumable electrodes, with additional cold-wire feed where necessary. The easy control of heat input makes TIG-welding the ideal welding method for fully orbital welding of tubes with specialist orbital welding heads, that incorporate a clamping device, a TIG electrode on an orbital travel device and a shielding gas chamber. Many different types of metal can be welded; high-strength, high-temperature and corrosion-resistant steels, unalloyed and low-alloyed carbon steels, nickel alloys, titanium, copper, aluminum and associated alloys. Carried out in an inert atmosphere, this controlled technique produces results that are extremely clean, have low particle counts and are free from unwanted spatter. This enables the highest demands to be met regarding the mechanical and optical properties of a weld seam.

Tube Diameters

Due to the precision of orbital TIG welding, even the smallest standard tube diameters from 1.6 millimetres can be processed. On a larger scale, pipes with diameters up to 170mm and walls up to 3.5mm thick can be joined using closed chamber weld heads. These weld heads allow the torch to be positioned very precisely and ensure that the pipe is held securely. The inert gas atmosphere in the closed chamber prevents heat from tinting, even with the most sensitive of materials. For tube diameters between 8 and 275mm, it is possible to use more manageable open welding heads (except for high purity applications). A flexible hose system is used to supply the welding head with power, inert gas, cooling water and filler wire where required. The need for filler wire during the welding process depends on the type of welding task; thicker tube walls and difficult-to-control parent materials require the use of additional material, whereas thin-walled tubes can be welded without extra wire. In order to create high quality weld seams it is essential that tube ends are carefully prepared with the edges of the workpieces being free of scale and impurities. For thinner-walled tubes up to medium diameters, a simple right-angled saw cut is often sufficient. For thicker tube walls it is necessary to prepare the edges more carefully, for example using a U-groove cross-section.

Business Sectors and Markets

Hot wire narrow gap welding in the production of power stations

Owing to its ability to realise high purity results, orbital welding found its place in the production of clean-room components for the semiconductor industry. Its application has now expanded to the construction of pipework and equipment for diverse industries like food processing, pharmaceuticals, chemical engineering, automotive engineering, biotechnology, shipbuilding and aerospace. Automated orbital TIG welding is also used in the construction of power stations, (thermal power plants). The construction materials used must be able to withstand the enormous mechanical loads produced by the high pressures and temperatures created by the media carried in the tubes. Notches, pores and inclusions in the weld seams must be avoided at all costs, as these create weak points that can lead to subsequent formation of cracks. These in turn can have serious consequences in terms of component failure. This means that tubes are often made from nickel-based materials with walls up to 200mm thick. One manufacturer has developed an orbital narrow gap welding system with hot-wire feed specifically for this purpose, which uses running gear that moves on a guide ring fixed around the tube. This new variant has created a lot of interest in the sector, with the worldwide boom in power station construction fueling the never-ending search for increasingly productive manufacturing methods using new types of high-temperature steels.

Conclusion

Along with the current methods of using TIG cold and hot wire welding, there has also been steady progress in the development of MIG/MAG/FCAW welding which allows a whole range of new applications in a variety of industries including aerospace, medical, automotive and more. Orbital welding can provide reliable welding of reproducible quality using wide-ranging techniques and differing types of technique. This can be performed to a high standard even when using unusual materials, thick walls, small tube diameters and even within a difficult working environment. The cost of orbital welding equipment is 5-10 times the initial capital cost required in conventional welding equipment but the productivity is also significantly higher than conventional TIG (2-3 times).

Pattern Welding

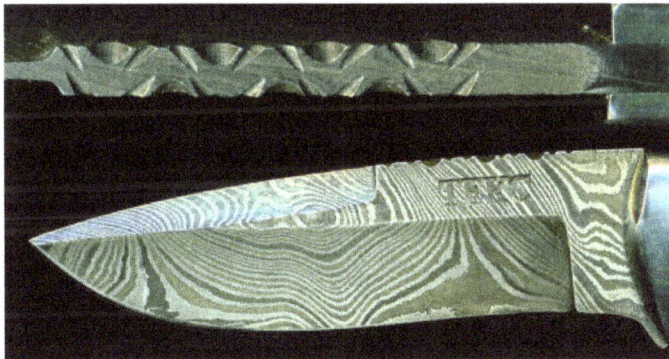

A high resolution image of a modern pattern welded knife blade, showing the dramatic patterning on the side below, and the layering of the steel in the spine above. Acid etching darkens the 1080 plain carbon steel more than it does the 15N20 low alloy nickel steel, producing alternating bands of light and dark on the surface.

Pattern welding is the practice in sword and knife making of forming a blade of several metal pieces of differing composition that are forge-welded together and twisted and manipulated to form a pattern. Often called (Modern) Damascus steel, blades forged in this manner often display bands of slightly different patterning along their entire length. These bands can be highlighted for cosmetic purposes by proper polishing or acid etching. Pattern welding was an outgrowth of laminated or piled steel, a similar technique used to combine steels of different carbon contents, providing a desired mix of hardness and toughness. Although modern steelmaking processes negate the need to blend different steels, pattern welded steel is still used by custom knifemakers for the cosmetic effects it produces.

History

Pattern-welded 19th century Moro (Philippine) barung sword

Close-up view of the blade of the same Moro barung

Pattern welding developed out of the necessarily complex process of making blades that were both hard and tough from the erratic and unsuitable output from early iron smelting in bloomeries. The bloomery does not generate temperatures high enough to melt iron and steel, but instead reduces the iron oxide ore into particles of pure iron, which then weld into a mass of sponge iron, consisting of lumps of impurities in a matrix of relatively pure iron, which is too soft to make a good blade. Carburizing thin iron bars or plates forms a layer of harder, high carbon steel on the surface, and early bladesmiths would forge these bars or plates together to form relatively homogeneous bars of steel. This laminating process, in which different types of steels together produces patterns that can be seen in the surface of the finished blade, forms the basis for pattern welding.

Pattern Welding in Europe

By the 2nd and 3rd century AD, the Celts commonly used pattern welding for decoration in addition to structural reasons. The technique involves folding and forging alternating layers of steel into rods, then twisting the steel to form complex patterns when forged into a blade. By the 6th and 7th centuries, pattern welding had reached a level where thin layers of patterned steel were being overlaid onto a soft iron core, making the swords far better as the iron gave them a flexible and springy core that would take any shock from sword blows to stop the blade bending or snapping. By the end of the Viking era, pattern welding fell out of use in Europe

During the Middle ages, Damascus steel was produced in India and brought back to Europe. The similarities in the markings led many to believe it was the same process being used, and pattern welding was revived by European smiths who were attempting to duplicate the Damascus steel. While the methods used by Damascus smiths to produce their blades was lost, recent efforts by metallurgists and bladesmiths (such as Verhoeven and Pendray) to reproduce steel with identical characteristics have yielded a process that does not involve pattern welding. However even these attempts have not been a huge success.

A similar technique was also employed by Scandinavian Medieval swordsmiths. The Mora knife is today manufactured with a similar technique. Today the traditional crucible steel is seldom used, but the high carbon steel is usually tool steel or stainless steel.

Modern Decorative Use

The ancient swordmakers exploited the aesthetic qualities of pattern welded steel. The Vikings, in particular, were fond of twisting bars of steel around each other, welding the bars together by hammering and then repeating the process with the resulting bars, to create complex patterns in the final steel bar. Two bars twisted in opposite directions created the common chevron pattern. Often, the center of the blade was a core of soft steel, and the edges were solid high carbon steel, similar to the laminates of the Japanese.

The American Bladesmith Society's Master Smith test, for example, requires a 300 layer blade to be forged. Large numbers of layers are generally produced by folding, where a small number of layers are welded together, then the blank is cut in half, stacked, and welded again, with each operation doubling the number of layers. Starting with just two layers, eight folding operations will yield 512 layers in the blank. A blade ground from such a blank will show a grain much like an object cut from a block of wood, with similar random variations in pattern. Some manufactured objects can be re-purposed into pattern welded blanks. "Cable Damascus", forged from high carbon multi-strand cable, is a popular item for bladesmiths to produce, producing a finely grained, twisted pattern, while chainsaw chains produce a pattern of randomly positioned blobs of color.

Some modern bladesmiths have taken pattern welding to new heights, with elaborate applications of traditional pattern welding techniques, as well as with new technology. A layered billet of steel rods with the blade blank cut perpendicular to the layers can also produce some spectacular patterns, including mosaics or even writing. Powder metallurgy allows alloys that would not normally be compatible to be combined into solid bars. Different treatments of the steel after it is ground and polished, such as bluing, etching, or various other chemical surface treatments that react differently to the different metals used can create bright, high-contrast finishes on the steel. Some master smiths go as far as to use techniques such as electrical discharge machining to cut interlocking patterns out of different steels, fit them together, then weld the resulting assembly into a solid block of steel.

Plastic Welding: An Integrated Study

Plastic welding as a process has three stages, namely surface preparation, application of heat and pressure and cooling. Along with plastic welding, friction welding, heat fusion, spin welding and ultrasonic welding have been elucidated in the following text. This section will provide an integrated understanding of plastic welding.

Plastic Welding

Plastic welding : welding for semi-finished plastic materials is described in ISO 472 as a process of uniting softened surfaces of materials, generally with the aid of heat (except solvent welding). Welding of thermoplastics is accomplished in three sequential stages, namely surface preparation, application of heat and pressure, and cooling. Numerous welding methods have been developed for the joining of semifinished plastic materials. Based on the mechanism of heat generation at the welding interface, welding methods for thermoplastics can be classified as external and internal heating methods, as shown in Fig 1.

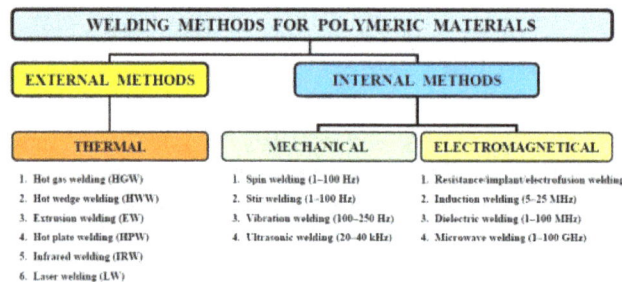

Fig. 1. Classification of welding methods for semi-finished polymeric materials.

On the other hand, production of a good quality weld can not only depend on the welding methods, but also weldability of base materials. Therefore, the evaluation of weldability is of critical importance before welding operation for plastics.

Welding Techniques

A number of techniques are used for welding of semi-finished plastic products as given below:

Hot Gas Welding

Hot gas welding, also known as *hot air welding*, is a plastic welding technique using heat. A specially designed heat gun, called a *hot air welder*, produces a jet of hot air that softens both the parts to be joined and a plastic filler rod, all of which must be of the same or a very similar plastic. (Welding PVC to acrylic is an exception to this rule.)

Hot air/gas welding is a common fabrication technique for manufacturing smaller items such as chemical tanks, water tanks, heat exchangers, and plumbing fittings.

In the case of webs and films a filler rod may not be used. Two sheets of plastic are heated via a hot gas (or a heating element) and then rolled together. This is a quick welding process and can be performed continuously.

Welding Rod

A plastic welding rod, also known as a *thermoplastic welding rod*, is a rod with circular or triangular cross-section used to bind two pieces of plastic together. They are available in a wide range of colors to match the base material's color. Spooled plastic welding rod is known as "spline".

An important aspect of plastic welding rod design and manufacture is the porosity of the material. A high porosity will lead to air bubbles (known as *voids*) in the rods, which decrease the quality of the welding. The highest quality of plastic welding rods are therefore those with zero porosity, which are called *voidless*.

Heat Sealing

Heat sealing is the process of sealing one thermoplastic to another similar thermoplastic using heat and pressure. The direct contact method of heat sealing utilizes a constantly heated die or sealing bar to apply heat to a specific contact area or path to seal or weld the thermoplastics together. Heat sealing is used for many applications, including heat seal connectors, thermally activated adhesives and film or foil sealing. Common applications for the heat sealing process: Heat seal connectors are used to join LCDs to PCBs in many consumer electronics, as well as in medical and telecommunication devices. Heat sealing of products with thermal adhesives is used to hold clear display screens onto consumer electronic products and for other sealed thermo-plastic assemblies or devices where heat staking or ultrasonic welding is not an option due to part design requirements or other assembly considerations. Heat sealing also is used in the manufacturing of bloodtest film and filter media for the blood, virus and many other test strip devices used in the medical field today. Laminate foils and films often are heat sealed over the top of thermoplastic medical trays, Microtiter (microwell) plates, bottles and containers to seal and/or prevent contamination for medical test devices, sample collection trays and containers used for food products. Medical and the Food Industries manufacturing Bag or flexible containers use heat sealing for either perimeter welding of the plastic material of the bags and/or for sealing ports and tubes into the bags. A variety of heat sealers are available to join thermoplastic materials such as plastic films: Hot bar sealer, Impulse sealer, etc.

Freehand Welding

With freehand welding, the jet of hot air (or inert gas) from the welder is played on the weld area and the tip of the weld rod at the same time. As the rod softens, it is pushed into the joint and fuses to the parts. This process is slower than most others, but it can be used in almost any situation.

Speed Tip Welding

With speed welding, the plastic welder, similar to a soldering iron in appearance and wattage, is fitted with a feed tube for the plastic weld rod. The speed tip heats the rod and the substrate, while at the same time it presses the molten weld rod into position. A bead of softened plastic is laid into the joint, and the parts and weld rod fuse. With some types of plastic such as polypropylene, the melted welding rod must be "mixed" with the semi-melted base material being fabricated or repaired. These welding techniques have been improved over time and have been utilized for over 50 years by professional plastic fabricators and repairers internationally. Speed tip welding method is a much faster welding technique and with practice can be used in tight corners. A version of the speed tip "gun" is essentially a soldering iron with a broad, flat tip that can be used to melt the weld joint and filler material to create a bond.

Extrusion Welding

Extrusion welding allows the application of bigger welds in a single weld pass. It is the preferred technique for joining material over 6 mm thick. Welding rod is drawn into a miniature hand held plastic extruder, plasticized, and forced out of the extruder against the parts being joined, which are softened with a jet of hot air to allow bonding to take place.

Contact Welding

This is the same as spot welding except that heat is supplied with thermal conduction of the pincher tips instead of electrical conduction. Two plastic parts are brought together where heated tips pinch them, melting and joining the parts in the process.

Hot Plate Welding

Related to contact welding, this technique is used to weld larger parts, or parts that have a complex weld joint geometry. The two parts to be welded are placed in the tooling attached to the two opposing platens of a press. A hot plate, with a shape that matches the weld joint geometry of the parts to be welded, is moved in position between the two parts. The two opposing platens move the parts into contact with the hot plate until the heat softens the interfaces to the melting point of the plastic. When this condition is achieved the hot plate is removed, and the parts are pressed together and held until the weld joint cools and re-solidifies to create a permanent bond.

Hot-plate welding equipment is typically controlled pneumatically, hydraulically, or electrically with servo motors.

This process is used to weld automotive under hood components, automotive interior trim components, medical filtration devices, consumer appliance components, and other car interior components.

High Frequency Welding

Certain plastics with chemical dipoles, such as PVC, polyamides (PA) and acetates can be heated with high frequency electromagnetic waves. High frequency welding uses this property to soften the plastics for joining. The heating can be localized, and the process can be continuous. Also known as Dielectric Sealing, R.F. (Radio Frequency) Heat Sealing.

Radio frequency welding is a very mature technology that has been around since the 1940s. Two pieces of material are placed on a table press that applies pressure to both surface areas. Dies are used to direct the welding process. When the press comes together, high frequency waves (usually 27.120 MHz) are passed through the small area between the die and the table where the weld takes place. This high frequency (radio frequency) field causes the molecules in certain materials to move and get hot, and the combination of this heat under pressure causes the weld to take the shape of the die. RF welding is fast. This type of welding is used to connect polymer films used in a variety of industries where a strong consistent leak-proof seal is required. In the fabrics industry, RF is most often used to weld PVC and polyurethane (PU) coated fabrics. This is a very consistent method of welding.

The most common materials used in RF welding are PVC and polyurethane. It is also possible to weld other polymers such as Nylon, PET, PEVA, EVA and some ABS plastics. Exercise caution when welding urethane as it has been known to give off cyanide gasses when melting.

Induction Welding

When an electrical insulator, like a plastic, is embedded with a material having high electrical conductivity, like metals or carbon fibers, induction welding can be performed. The welding apparatus contains an induction coil that is energised with a radio-frequency electric current. This generates an electromagnetic field that acts on either an electrically conductive or a ferromagnetic workpiece. In an electrically conductive workpiece, the main heating effect is resistive heating, which is due to induced currents called eddy currents. Induction welding of carbon fiber reinforced thermoplastic materials is a technology commonly used in for instance the aerospace industry.

In a ferromagnetic workpiece, plastics can be induction-welded by formulating them with metallic or ferromagnetic compounds, called susceptors. These susceptors absorb electromagnetic energy from an induction coil, become hot, and lose their heat energy to the surrounding material by thermal conduction.

Injection Welding

Injection welding is similar/identical to extrusion welding, except, using certain tips on the hand-held welder, one can insert the tip into plastic defect holes of various sizes and patch them from the inside out. The advantage is that no access is needed to the rear of the defect hole. The alternative is a patch, except that the patch can not be sanded flush with the original surrounding plastic to the same thickness. PE and PP are most suitable for this type of process. The Drader injectiweld is an example of such tool.

Ultrasonic Welding

In ultrasonic welding, high frequency (15 kHz to 40 kHz) low amplitude vibration is used to create heat by way of friction between the materials to be joined. The interface of the two parts is specially designed to concentrate the energy for the maximum weld strength. Ultrasonic can be used on almost all plastic material. It is the fastest heat sealing technology available.

Friction Welding

In friction welding, the two parts to be assembled are rubbed together at a lower frequency (typically 100–300 Hz) and higher amplitude (typically 1 to 2 mm (0.039 to 0.079 in)) than ultrasonic welding. The friction caused by the motion combined with the clamping pressure between the two parts creates the heat which begins to melt the contact areas between the two parts. At this point, the plasticized materials begin to form layers that intertwine with one another, which therefore results in a strong weld. At the completion of the vibration motion, the parts remain held together until the weld joint cools and the melted plastic re-solidifies. The friction movement can be linear or orbital, and the joint design of the two parts has to allow this movement.

Spin Welding

Spin welding is a particular form of frictional welding. With this process, one component with a round weld joint is held stationary, while a mating component is rotated at high speed and pressed against the stationary component. The rotational friction between the two components generates heat. Once the joining surfaces reach a semi-molten state, the spinning component is stopped abruptly. Force on the two components is maintained until the weld joint cools and re-solidifies. This is a common way of producing low- and medium-duty plastic wheels, e.g., for toys, shopping carts, recycling bins, etc. This process is also used to weld various port openings into automotive under hood components.

Laser Welding

This technique requires one part to be transmissive to a laser beam and either the other part absorptive or a coating at the interface to be absorptive to the beam. The two parts are put under pressure while the laser beam moves along the joining line. The beam passes through the first part and is absorbed by the other one or the coating to generate enough heat to soften the interface creating a permanent weld.

Semiconductor diode lasers are typically used in plastic welding. Wavelengths in the range of 808 nm to 980 nm can be used to join various plastic material combinations. Power levels from less than 1W to 100W are needed depending on the materials, thickness and desired process speed.

Diode laser systems have the following advantages in joining of plastic materials:

- Cleaner than adhesive bonding
- No micro-nozzles to get clogged
- No liquid or fumes to affect surface finish
- No consumables
- Higher throughput
- Can access work-piece in challenging geometry
- High level of process control

Requirements for high strength joints include:

- Adequate transmission through upper layer

- Absorption by lower layer

- Material compatibility – wetting

- Good joint design – clamping pressure, joint area

- Lower power density

A sample list of materials that can be joined include:

- Polypropylene

- Polycarbonate

- Acrylic

- Nylon

- ABS

Specific applications include sealing / welding / joining of: catheter bags, medical containers, automobile remote control keys, heart pacemaker casings, syringe tamper evident joints, headlight or tail-light assemblies, pump housings, and cellular phone parts.

Transparent Laser Plastic Welding

New fiber laser technology allows for the output of higher laser wavelengths, with the best results typically around 2,000 nm, significantly higher than the average 808 nm to 1064 nm diode laser used for traditional laser plastic welding. Because these higher wavelengths are more readily absorbed by thermoplastics than the infra-red radiation of traditional plastic welding, it is possible to weld two clear polymers without any colorants or absorbing additives. Common Applications will mostly fall in the medical industry for devices like catheters and microfluidic devices. The heavy use of transparent plastics, especially flexible polymers like TPU, TPE and PVC, in the medical device industry makes transparent laser welding a natural fit. Also, the process requires no laser absorbing additives or colorants making testing and meeting biocompatibility requirements significantly easier.

Solvent Welding

In solvent welding, a solvent is applied which can temporarily dissolve the polymer at room temperature. When this occurs, the polymer chains are free to move in the liquid and can mingle with other similarly dissolved chains in the other component. Given sufficient time, the solvent will permeate through the polymer and out into the environment, so that the chains lose their mobility. This leaves a solid mass of entangled polymer chains which constitutes a solvent weld.

This technique is commonly used for connecting PVC and ABS pipe, as in household plumbing. The "gluing" together of plastic (polycarbonate, polystyrene or ABS) models is also a solvent welding process.

Dichloromethane (methylene chloride), which is obtainable in paint stripper, can solvent weld polycarbonate and polymethylmethacrylate. Dichloromethane chemically welds certain plastics; for example, it is used to seal the casing of electric meters. It is also a component – along with tetrahydrofuran – of the solvent used to weld plumbing.

Friction Welding

Friction welding (FRW) is a solid-state welding process that generates heat through mechanical friction between workpieces in relative motion to one another, with the addition of a lateral force called "upset" to plastically displace and fuse the materials. Because no melting occurs, friction welding is not a fusion welding process in the traditional sense, but more of a forge welding technique. Friction welding is used with metals and thermoplastics in a wide variety of aviation and automotive applications.

Benefits

The combination of fast joining times (on the order of a few seconds), and direct heat input at the weld interface, yields relatively small heat-affected zones. Friction welding techniques are generally melt-free, which mitigates grain growth in engineered materials, such as high-strength heat-treated steels. Another advantage is that the motion tends to "clean" the surface between the materials being welded, which means they can be joined with less preparation. During the welding process, depending on the method being used, small pieces of the plastic or metal will be forced out of the working mass (flash). It is believed that the flash carries away debris and dirt.

Another advantage of friction welding is that it allows dissimilar materials to be joined. This is particularly useful in aerospace, where it is used to join lightweight aluminum stock to high-strength steels. Normally the wide difference in melting points of the two materials would make it impossible to weld using traditional techniques, and would require some sort of mechanical connection. Friction welding provides a "full strength" bond with no additional weight. Other common uses for these sorts of bi-metal joins is in the nuclear industry, where copper-steel joints are common in the reactor cooling systems; and in the transport of cryogenic fluids, where friction welding has been used to join aluminum alloys to stainless steels and high-nickel-alloy materials for cryogenic-fluid piping and containment vessels.

Friction welding is also used with thermoplastics, which act in a fashion analogous to metals under heat and pressure. The heat and pressure used on these materials is much lower than metals, but the technique can be used to join metals to plastics with the metal interface being machined. For instance, the technique can be used to join eyeglass frames to the pins in their hinges. The lower energies and pressures used allows for a wider variety of techniques to be used.

History

Friction welding was first developed in the Soviet Union, with first experiments taking place in 1956. The American companies Caterpillar, Rockwell International, and American Manufacturing

Foundry all developed machines for this process. Patents were also issued throughout Europe and the former Soviet Union. The most extensive historical records are kept with the American Welding Society.

Metal Techniques

Spin Welding

Spin welding systems consist of two chucks for holding the materials to be welded, one of which is fixed and the other rotating. Before welding one of the work pieces is attached to the rotating chuck along with a flywheel of a given weight. The piece is then spun up to a high rate of rotation to store the required energy in the flywheel. Once spinning at the proper speed, the motor is removed and the pieces forced together under pressure. The force is kept on the pieces after the spinning stops to allow the weld to "set".

In Inertia Friction Welding the drive motor is disengaged, and the work pieces are forced together by a friction welding force. The kinetic energy stored in the rotating flywheel is dissipated as heat at the weld interface as the flywheel speed decreases.

In Direct Drive Friction welding the drive motor and chuck are connected. The drive motor is continually driving the chuck during the heating stages. Usually a clutch is used to disconnect the drive motor from the chuck and a brake is then used to stop the chuck.

Linear Friction Welding

Linear friction welding (*LFW*) is similar to spin welding except that the moving chuck oscillates laterally instead of spinning. The speeds are much lower in general, which requires the pieces to be kept under pressure at all times. This also requires the parts to have a high shear strength. Linear friction welding requires more complex machinery than spin welding, but has the advantage that parts of any shape can be joined, as opposed to parts with a circular meeting point. Another advantage is that in many instances quality of joint is better than that obtained using rotating technique.

Friction Surfacing

Friction surfacing is a process derived from friction welding where a coating material is applied to a substrate. A rod composed of the coating material (called a mechtrode) is rotated under pressure, generating a plasticised layer in the rod at the interface with the substrate. By moving a substrate across the face of the rotating rod a plasticised layer is deposited between 0.2–2.5 millimetres (0.0079–0.0984 in) thick depending on mechtrode diameter and coating material.

Thermoplastic Techniques

Linear Vibration Welding

In *linear vibration welding* the materials are placed in contact and put under pressure. An external vibration force is then applied to slip the pieces relative to each other, perpendicular to the pressure being applied. The parts are vibrated through a relatively small displacement known as the amplitude, typically between 1.0 and 1.8 mm, for a frequency of vibration of 200 Hz (high frequency), or 2–4 mm at 100 Hz (low frequency), in the plane of the joint. This technique is widely used in the automotive industry, among others. A minor modification is *angular friction welding*, which vibrates the materials by torquing them through a small angle.

Orbital Friction Welding

Orbital friction welding is similar to spin welding, but uses a more complex machine to produce an orbital motion in which the moving part rotates in a small circle, much smaller than the size of the joint as a whole.

Seizure Resistance

Friction welding may unintentionally occur at sliding surfaces like bearings. This happens in particular if the lubricating oil film between sliding surfaces becomes thinner than the surface roughness, which may be the case for low speed, low temperature, oil starvation, excessive clearance, low viscosity of the oil, high roughness of the surfaces.

The seizure resistance is the ability of a material to resist friction welding. It is a fundamental property of bearing surfaces and in general of sliding surfaces under load.

Heat Fusion

Heat fusion (sometimes called heat welding, butt welding or simply fusion) is a welding process used to join two different pieces of a thermoplastic. This process involves heating both pieces simultaneously and pressing them together. The two pieces then cool together and form a permanent bond. When done properly, the two pieces become indistinguishable from each other. Dissimilar plastics can result in improper bonding.

Applications

This process is commonly used in plastic pressure pipe systems to join a pipe and fitting together, or to join a length of pipe directly to another length of pipe. Generally, polyolefins (such as polypropylene, polyethylene, and polybutylene) are used for these applications.

Types

Butt welding is usually performed using one of several methods. The first, and most common, is *butt welding* or *butt fusion*, which is a type of hot plate welding. This technique involves heating two planed surfaces of thermoplastic material (typically polyethylene) against a heated surface. After a specified amount of time, the heating plate is removed and the two pieces are pressed together and allowed to cool under pressure, forming the desired bond. Butt welding outside of manufacturing is usually performed to join pipes.

The other major technique is *socket fusion*. It is distinguished from butt-welding by using custom-shaped and -sized heating plates rather than a basic flat surface. These heads allow for more surface contact, reducing the time needed to heat and fuse the pipe. Socket fusion joins pipe and fittings together, rather than simply joining pipe to pipe. It requires less pressure than butt-welding and is more commonly used on smaller sizes of pipe (4" or less). Socket welding has additional advantages of requiring less machinery and is more portable than the heavier equipment required for butt fusion.

A third method of thermoplastic welding is called *sidewall fusion*, or *saddle fusion*. Sidewall fusion is, like butt fusion and socket fusion, another process based on hot plate welding. Sidewall fusion differs from either socket, or butt fusion methods by performing fusion into the side of the pipe wall in a transverse orientation to the main pipe, rather than in line with the pipe. Sidewall fusion is typically employed in conjunction with either socket or butt fusion methods as a complementary process and many fusion machines designed for butt fusion are also equipped for sidewall fusion. Adaptor plates that match the outside diameter of the main pipe are applied to the heating plate to perform this type of fusion.

Another method used is referred to as *electrofusion*. Electrofusion is a method of joining HDPE and other plastic pipes with special fittings that have built-in resistive wire which is used to weld the joint together. The pipes to be joined are trimmed, cleaned, inserted into the electrofusion fitting (with a temporary clamp if required) and a voltage (typically 40V) is applied using a device called an electrofusion processor. The processor controls how much voltage is applied, and for how long, depending on the fitting in use. As current is applied to the resistive wire, the coils heat up and melt the inside of the fitting and the outside of the pipe wall which weld together producing a very strong homogeneous joint. The assembly is then left to cool for a specified time. The joints produced tend to be more reliable than threaded fittings sealed with O-rings.

Spin Welding

Spin welding is a friction welding technique used on thermoplastic materials, in which the parts to be welded are heated by friction. The heat may be generated by turning on a lathe, a drill press, or a milling machine, where one part is driven by the chuck, and the other is held stationary with the spinning part driven against it. This is continued until the heat of friction between the parts reaches a sufficient level for the parts to weld. The stationary part is then released to spin as well, while pressure is applied along the axis of rotation, holding the parts together as they cool.

Ultrasonic Welding

Ultrasonic welding is an industrial technique whereby high-frequency ultrasonic acoustic vibrations are locally applied to workpieces being held together under pressure to create a solid-state weld. It is commonly used for plastics, and especially for joining dissimilar materials. In ultrasonic welding, there are no connective bolts, nails, soldering materials, or adhesives necessary to bind the materials together.

Ultrasonic welding of thin metallic foils. The sonotrode is rotated along the weld seam.

History

Practical application of ultrasonic welding for rigid plastics was completed in the 1960s. At this point only hard plastics could be welded. The patent for the ultrasonic method for welding rigid thermoplastic parts was awarded to Robert Soloff and Seymour Linsley in 1965. Soloff, the founder of Sonics & Materials Inc., was a lab manager at Branson Instruments where thin plastic films were welded into bags and tubes using ultrasonic probes. He unintentionally moved the probe close to a plastic tape dispenser and the halves of the dispenser welded together. He realized that the probe did not need to be manually moved around the part but that the ultrasonic energy could travel through and around rigid plastics and weld an entire joint. He went on to develop the first ultrasonic press. The first application of this new technology was in the toy industry.

The first car made entirely out of plastic was assembled using ultrasonic welding in 1969. Even though plastic cars did not catch on, ultrasonic welding did. The automotive industry has used it regularly since the 1980s. It is now used for a multitude of applications.

Process

For joining complex injection molded thermoplastic parts, ultrasonic welding equipment can be easily customized to fit the exact specifications of the parts being welded. The parts are sandwiched between a fixed shaped nest (anvil) and a sonotrode (horn) connected to a transducer, and a ~20 kHz low-amplitude acoustic vibration is emitted. (Note: Common frequencies used in ultrasonic welding of thermoplastics are 15 kHz, 20 kHz, 30 kHz, 35 kHz, 40 kHz and 70 kHz). When welding plastics, the interface of the two parts is specially designed to concentrate the melting process. One of the materials usually has a spiked energy director which contacts the second plastic part. The ultrasonic energy melts the point contact between the parts, creating a joint. This process is a good automated alternative to glue, screws or snap-fit designs. It is typically used with small parts (e.g. cell phones, consumer electronics, disposable medical tools, toys, etc.) but it can be used on parts as large as a small automotive instrument cluster. Ultrasonics can also be used to weld metals, but are typically limited to small welds of thin, malleable metals, e.g. aluminum, copper, nickel. Ultrasonics would not be used in welding the chassis of an automobile or in welding pieces of a bicycle together, due to the power levels required.

Ultrasonic welding of thermoplastics causes local melting of the plastic due to absorption of vibrational energy along the joint to be welded. In metals, welding occurs due to high-pressure dispersion of surface oxides and local motion of the materials. Although there is heating, it is not enough to melt the base materials.

Ultrasonic welding can be used for both hard and soft plastics, such as semicrystalline plastics, and metals. Ultrasonic welding machines also have much more power now. The understanding of ultrasonic welding has increased with research and testing. The invention of more sophisticated and inexpensive equipment and increased demand for plastic and electronic components has led to a growing knowledge of the fundamental process. However, many aspects of ultrasonic welding still require more study, such as relating weld quality to process parameters. Ultrasonic welding continues to be a rapidly developing field.

Scientists from the Institute of Materials Science and Engineering (WKK) of University of Kaiserslautern, with the support from the German Research Foundation (Deutsche Forschungsgemeinschaft), have succeeded in proving that using ultrasonic welding processes can lead to highly durable bonds between light metals and Carbon-fiber-reinforced polymer (CFRP) sheets.

The benefits of ultrasonic welding are that it is much faster than conventional adhesives or solvents. The drying time is very quick, and the pieces do not need to remain in a jig for long periods of time waiting for the joint to dry or cure. The welding can easily be automated, making clean and precise joints; the site of the weld is very clean and rarely requires any touch-up work. The low thermal impact on the materials involved enables a greater number of materials to be welded together.

Components

All ultrasonic welding systems are composed of the same basic elements:

- A press to put the two parts to be assembled under pressure
- A nest or anvil where the parts are placed and allowing the high frequency vibration to be directed to the interfaces
- An ultrasonic stack composed of a converter or piezoelectric transducer, an optional booster and a sonotrode (US: Horn). All three elements of the stack are specifically tuned to resonate at the same exact ultrasonic frequency (Typically 20, 30, 35 or 40 kHz)
  - Converter: Converts the electrical signal into a mechanical vibration
  - Booster: Modifies the amplitude of the vibration. It is also used in standard systems to clamp the stack in the press.
  - Sonotrode: Applies the mechanical vibration to the parts to be welded.
- An electronic ultrasonic generator (US: Power supply) delivering a high power AC signal with frequency matching the resonance frequency of the stack.
- A controller controlling the movement of the press and the delivery of the ultrasonic energy.

Applications

The applications of ultrasonic welding are extensive and are found in many industries including electrical and computer, automotive and aerospace, medical, and packaging. Whether two items can be ultrasonically welded is determined by their thickness. If they are too thick this process will not join them. This is the main obstacle in the welding of metals. However, wires, microcircuit connections, sheet metal, foils, ribbons and meshes are often joined using ultrasonic welding. Ultrasonic welding is a very popular technique for bonding thermoplastics. It is fast and easily automated with weld times often below one second and there is no ventilation system required to remove heat or exhaust. This type of welding is often used to build assemblies that are too small, too complex, or too delicate for more common welding techniques.

Computer and Electrical Industries

In the electrical and computer industry ultrasonic welding is often used to join wired connections and to create connections in small, delicate circuits. Junctions of wire harnesses are often joined using ultrasonic welding. Wire harnesses are large groupings of wires used to distribute electrical signals and power. Electric motors, field coils, transformers and capacitors may also be assembled with ultrasonic welding. It is also often preferred in the assembly of storage media such as flash drives and computer disks because of the high volumes required. Ultrasonic welding of computer disks has been found to have cycle times of less than 300 ms.

One of the areas in which ultrasonic welding is most used and where new research and experimentation is centered is microcircuits. This process is ideal for microcircuits since it creates reliable bonds without introducing impurities or thermal distortion into components. Semiconductor devices, transistors and diodes are often connected by thin aluminum and gold wires using ultrasonic welding. It is also used for bonding wiring and ribbons as well as entire chips to microcircuits. An example of where microcircuits are used is in medical sensors used to monitor the human heart in bypass patients.

One difference between ultrasonic welding and traditional welding is the ability of ultrasonic welding to join dissimilar materials. The assembly of battery components is a good example of where this ability is utilized. When creating battery and fuel cell components, thin gauge copper, nickel and aluminum connections, foil layers and metal meshes are often ultrasonically welded together. Multiple layers of foil or mesh can often be applied in a single weld eliminating steps and costs.

Aerospace and Automotive Industries

For automobiles, ultrasonic welding tends to be used to assemble large plastic and electrical components such as instrument panels, door panels, lamps, air ducts, steering wheels, upholstery and engine components. As plastics have continued to replace other materials in the design and manufacture of automobiles, the assembly and joining of plastic components has increasingly become a critical issue. Some of the advantages for ultrasonic welding are low cycle times, automation, low capital costs, and flexibility. Also, ultrasonic welding does not damage surface finish, which is a crucial consideration for many car manufacturers, because the high-frequency vibrations prevent marks from being generated.

Ultrasonic welding is generally utilized in the aerospace industry when joining thin sheet gauge metals and other lightweight materials. Aluminum is a difficult metal to weld using traditional techniques because of its high thermal conductivity. However, it is one of the easier materials to weld using ultrasonic welding because it is a softer metal and thus a solid-state weld is simple to achieve. Since aluminum is so widely used in the aerospace industry, it follows that ultrasonic welding is an important manufacturing process. Also, with the advent of new composite materials, ultrasonic welding is becoming even more prevalent. It has been used in the bonding of the popular composite material carbon fiber. Numerous studies have been done to find the optimum parameters that will produce quality welds for this material.

Medical Industry

In the medical industry ultrasonic welding is often used because it does not introduce contaminants or degradation into the weld and the machines can be specialized for use in clean rooms. The process can also be highly automated, provides strict control over dimensional tolerances and does not interfere with the biocompatibility of parts. Therefore, it increases part quality and decreases production costs. Items such as arterial filters, anesthesia filters, blood filters, IV catheters, dialysis tubes, pipettes, cardiometry reservoirs, blood/gas filters, face masks and IV spike/filters can all be made using ultrasonic welding. Another important application in the medical industry for ultrasonic welding is textiles. Items like hospital gowns, sterile garments, masks, transdermal patches and textiles for clean rooms can be sealed and sewn using ultrasonic welding. This prevents contamination and dust production and reduces the risk of infection.

Packaging Industry

Butane lighter

Packaging is an application where ultrasonic welding is often used. Many common items are either created or packaged using ultrasonic welding. Sealing containers, tubes and blister packs are common applications.

Ultrasonic welding is also applied in the packaging of dangerous materials such as explosives, fireworks and other reactive chemicals. These items tend to require hermetic sealing but cannot be subjected to high temperatures. One example is a butane lighter. This container weld must be able to withstand high pressure and stress and must be airtight to contain the butane. Another example is the packaging of ammunition and propellants. These packages must be able to withstand high pressure and stress to protect the consumer from the contents. When sealing hazardous materials, safety is a primary concern.

The food industry finds ultrasonic welding preferable to traditional joining techniques because it is fast, sanitary and can produce hermetic seals. Milk and juice containers are examples of products often sealed using ultrasonic welding. The paper parts to be sealed are coated with plastic, generally polypropylene or polyethylene, and then welded together to create an airtight seal. The main obstacle to overcome in this process is the setting of the parameters. For example, if over-welding occurs then the concentration of plastic in the weld zone may be too low and cause the seal to break. If it is under-welded the seal is incomplete. Variations in the thicknesses of materials can cause variations in weld quality. Some other food items sealed using ultrasonic welding include candy bar wrappers, frozen food packages and beverage containers.

Safety

Ultrasonic welding machines, like most industrial equipment, pose the risk of some hazards. These include exposure to high heat levels and voltages. This equipment should be operated using the safety guidelines provided by the manufacturer to avoid injury. For instance, operators must never place hands or arms near the welding tip when the machine is activated. Also, operators should be provided with hearing protection and safety glasses. Operators should be informed of the OSHA regulations for the ultrasonic welding equipment and these regulations should be enforced.

Ultrasonic welding machines require routine maintenance and inspection. Panel doors, housing covers and protective guards may need to be removed for maintenance. This should be done when the power to the equipment is off and only by the trained professional servicing the machine.

Since this is an ultrasonic process it would seem that sound would not be an issue. However, sub-harmonic vibrations, which can create annoying audible noise, may be caused in larger parts near the machine due to the ultrasonic welding frequency. This noise can be damped by clamping these large parts at one or more locations. Also, high-powered welders with frequencies of 15 kHz and 20 kHz typically emit a potentially damaging high-pitched squeal in the range of human hearing. Shielding this radiating sound can be done using an acoustic enclosure. There are hearing and safety concerns with ultrasonic welding that are important to consider, but generally they are comparable to those of other welding techniques.

Brazing and Soldering: An Integrated Study

Brazing is done to join two metals; metal is melted and filled into the joint that connects the two metals. There is a difference between welding and brazing, as welding does not involve melting the work pieces. This chapter also explains to the reader the concepts of soldering, dip soldering, ultrasonic soldering and wave soldering.

Brazing

Brazing is a metal-joining process in which two or more metal items are joined together by melting and flowing a filler metal into the joint, the filler metal having a lower melting point than the adjoining metal.

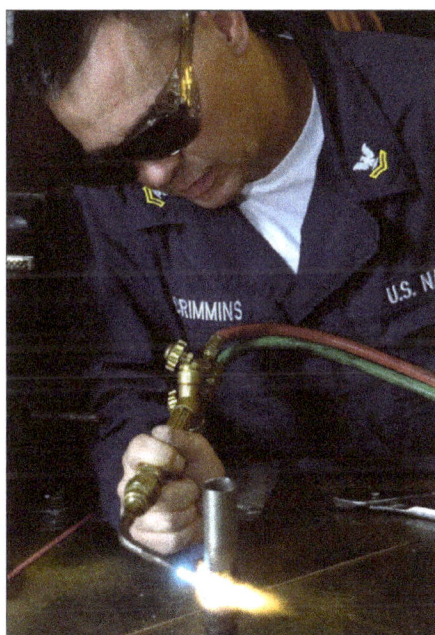

Brazing practice

Brazing differs from welding in that it does not involve melting the work pieces and from soldering in using higher temperatures for a similar process, while also requiring much more closely fitted parts than when soldering. The filler metal flows into the gap between close-fitting parts by capillary action. The filler metal is brought slightly above its melting (liquidus) temperature while protected by a suitable atmosphere, usually a flux. It then flows over the base metal (known as wetting) and is then cooled to join the work pieces together. It is similar to soldering, except for the use of higher temperatures. A major advantage of brazing is the ability to join the same or different metals with considerable strength.

Fundamentals

High-quality brazed joints require that parts be closely fitted, and the base metals exceptionally clean and free of oxides. In most cases, joint clearances of 0.03 to 0.08 mm (0.0012 to 0.0031 in) are recommended for the best capillary action and joint strength. However, in some brazing operations it is not uncommon to have joint clearances around 0.6 mm (0.024 in). Cleanliness of the brazing surfaces is also important, as any contamination can cause poor wetting (flow). The two main methods for cleaning parts, prior to brazing, are chemical cleaning and abrasive or mechanical cleaning. In the case of mechanical cleaning, it is important to maintain the proper surface roughness as wetting on a rough surface occurs much more readily than on a smooth surface of the same geometry.

Another consideration is the effect of temperature and time on the quality of brazed joints. As the temperature of the braze alloy is increased, the alloying and wetting action of the filler metal increases as well. In general, the brazing temperature selected must be above the melting point of the filler metal. However, several factors influence the joint designer's temperature selection. The best temperature is usually selected to:

- Be the lowest possible braze temperature

- Minimize any heat effects on the assembly

- Minimize filler metal/base metal interaction

- Maximize the life of any fixtures or jigs used

In some cases, a worker may select a higher temperature to accommodate other factors in the design (e.g., to allow use of a different filler metal, or to control metallurgical effects, or to sufficiently remove surface contamination). The effect of time on the brazed joint primarily affects the extent to which these effects are present. In general, however, most production processes are selected to minimize brazing time and associated costs. This is not always the case, however, since in some non-production settings, time and cost are secondary to other joint attributes (e.g., strength, appearance).

Flux

Unless brazing operations are contained within an inert or reducing atmosphere environment (i.e. a vacuum furnace), a flux such as borax is required to prevent oxides from forming while the metal is heated. The flux also serves the purpose of cleaning any contamination left on the brazing surfaces. Flux can be applied in any number of forms including flux paste, liquid, powder or pre-made brazing pastes that combine flux with filler metal powder. Flux can also be applied using brazing rods with a coating of flux, or a flux core. In either case, the flux flows into the joint when applied to the heated joint and is displaced by the molten filler metal entering the joint. Excess flux should be removed when the cycle is completed because flux left in the joint can lead to corrosion, impede joint inspection, and prevent further surface finishing operations. Phosphorus-containing brazing alloys can be self-fluxing when joining copper to copper. Fluxes are generally selected based on their performance on particular base metals. To be effective, the flux must be chemically compatible with both the base metal and the filler metal being used. Self-fluxing phosphorus filler alloys produce brittle phosphides if used on iron or nickel. As a general rule, longer brazing cycles should use less active fluxes than short brazing operations.

Filler Materials

A variety of alloys are used as filler metals for brazing depending on the intended use or application method. In general, braze alloys are made up of 3 or more metals to form an alloy with the desired properties. The filler metal for a particular application is chosen based on its ability to: wet the base metals, withstand the service conditions required, and melt at a lower temperature than the base metals or at a very specific temperature.

Braze alloy is generally available as rod, ribbon, powder, paste, cream, wire and preforms (such as stamped washers). Depending on the application, the filler material can be pre-placed at the desired location or applied during the heating cycle. For manual brazing, wire and rod forms are generally used as they are the easiest to apply while heating. In the case of furnace brazing, alloy is usually placed beforehand since the process is usually highly automated. Some of the more common types of filler metals used are

- Aluminum-silicon

- Copper

- Copper-silver

- Copper-zinc (brass)

- Copper-tin (bronze)

- Gold-silver

- Nickel alloy

- Silver

- Amorphous brazing foil using nickel, iron, copper, silicon, boron, phosphorus, etc.

Atmosphere

As brazing work requires high temperatures, oxidation of the metal surface occurs in an oxygen-containing atmosphere. This may necessitate the use of an atmospheric environment other than air. The commonly used atmospheres are

- Air: Simple and economical. Many materials susceptible to oxidation and buildup of scale. Acid cleaning bath or mechanical cleaning can be used to remove the oxidation after work. Flux counteracts the oxidation, but may weaken the joint.

- Combusted fuel gas (low hydrogen, AWS type 1, "exothermic generated atmospheres"): 87% N_2, 11–12% CO_2, 5-1% CO, 5-1% H_2. For silver, copper-phosphorus and copper-zinc filler metals. For brazing copper and brass.

- Combusted fuel gas (decarburizing, AWS type 2, "endothermic generated atmospheres"): 70–71% N_2, 5–6% CO_2, 9–10% CO, 14–15% H_2. For copper, silver, copper-phosphorus and copper-zinc filler metals. For brazing copper, brass, nickel alloys, Monel, medium carbon steels.

- Combusted fuel gas (dried, AWS type 3, "endothermic generated atmospheres"): 73–75% N_2, 10–11% CO, 15–16% H_2. For copper, silver, copper-phosphorus and copper-zinc filler metals. For brazing copper, brass, low-nickel alloys, Monel, medium and high carbon steels.

- Combusted fuel gas (dried, decarburizing, AWS type 4): 41–45% N_2, 17–19% CO, 38–40% H_2. For copper, silver, copper-phosphorus and copper-zinc filler metals. For brazing copper, brass, low-nickel alloys, medium and high carbon steels.

- Ammonia (AWS type 5, also called forming gas): Dissociated ammonia (75% hydrogen, 25% nitrogen) can be used for many types of brazing and annealing. Inexpensive. For copper, silver, nickel, copper-phosphorus and copper-zinc filler metals. For brazing copper, brass, nickel alloys, Monel, medium and high carbon steels and chromium alloys.

- Nitrogen+hydrogen, cryogenic or purified (AWS type 6A): 70–99% N_2, 1–30% H_2. For copper, silver, nickel, copper-phosphorus and copper-zinc filler metals.

- Nitrogen+hydrogen+carbon monoxide, cryogenic or purified (AWS type 6B): 70–99% N_2, 2–20% H_2, 1–10% CO. For copper, silver, nickel, copper-phosphorus and copper-zinc filler metals. For brazing copper, brass, low-nickel alloys, medium and high carbon steels.

- Nitrogen, cryogenic or purified (AWS type 6C): Non-oxidizing, economical. At high temperatures can react with some metals, e.g. certain steels, forming nitrides. For copper, silver, nickel, copper-phosphorus and copper-zinc filler metals. For brazing copper, brass, low-nickel alloys, Monel, medium and high carbon steels.

- Hydrogen (AWS type 7): Strong deoxidizer, highly thermally conductive. Can be used for copper brazing and annealing steel. May cause hydrogen embrittlement to some alloys. For copper, silver, nickel, copper-phosphorus and copper-zinc filler metals. For brazing copper, brass, nickel alloys, Monel, medium and high carbon steels and chromium alloys, cobalt alloys, tungsten alloys, and carbides.

- Inorganic vapors (various volatile fluorides, AWS type 8): Special purpose. Can be mixed with atmospheres AWS 1–5 to replace flux. Used for silver-brazing of brasses.

- Noble gas (usually argon, AWS type 9): Non-oxidizing, more expensive than nitrogen. Inert. Parts must be very clean, gas must be pure. For copper, silver, nickel, copper-phosphorus and copper-zinc filler metals. For brazing copper, brass, nickel alloys, Monel, medium and high carbon steels chromium alloys, titanium, zirconium, hafnium.

- Noble gas+hydrogen (AWS type 9A)

- Vacuum: Requires evacuating the work chamber. Expensive. Unsuitable (or requires special care) for metals with high vapor pressure, e.g. silver, zinc, phosphorus, cadmium, and manganese. Used for highest-quality joints, for e.g. aerospace applications.

Common Techniques

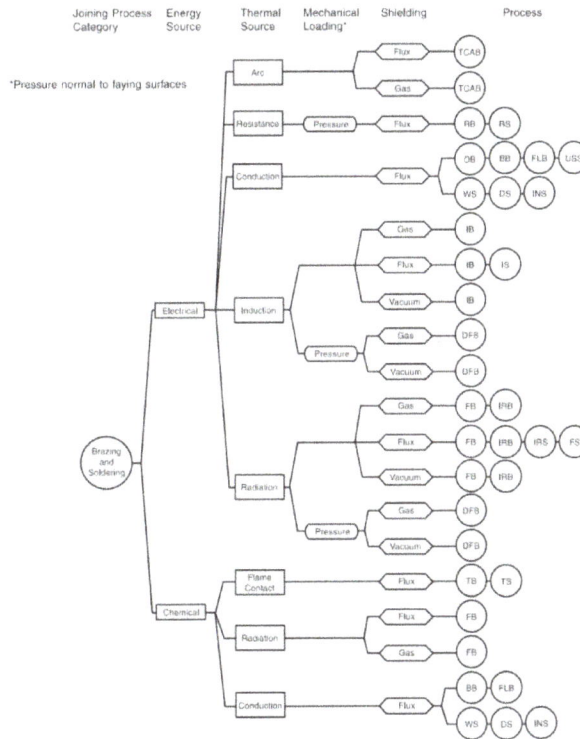

Brazing and soldering processes classification chart

Torch Brazing

Torch brazing is by far the most common method of mechanized brazing in use. It is best used in small production volumes or in specialized operations, and in some countries, it accounts for a majority of the brazing taking place. There are three main categories of torch brazing in use: manual, machine, and automatic torch brazing.

Manual torch brazing is a procedure where the heat is applied using a gas flame placed on or near the joint being brazed. The torch can either be hand held or held in a fixed position depending on whether the operation is completely manual or has some level of automation. Manual brazing is most commonly used on small production volumes or in applications where the part size or configuration makes other brazing methods impossible. The main drawback is the high labor cost associated with the method as well as the operator skill required to obtain quality brazed joints. The use of flux or self-fluxing material is required to prevent oxidation. Torch brazing of copper can be done without the use of flux if it is brazed with a torch using oxygen and hydrogen gas, rather than oxygen and other flammable gases.

Machine torch brazing is commonly used where a repetitive braze operation is being carried out. This method is a mix of both automated and manual operations with an operator often placing brazes material, flux and jigging parts while the machine mechanism carries out the actual braze. The advantage of this method is that it reduces the high labor and skill requirement of manual brazing. The use of flux is also required for this method as there is no protective atmosphere, and it is best suited to small to medium production volumes.

Automatic torch brazing is a method that almost eliminates the need for manual labor in the brazing operation, except for loading and unloading of the machine. The main advantages of this method are: a high production rate, uniform braze quality, and reduced operating cost. The equipment used is essentially the same as that used for Machine torch brazing, with the main difference being that the machinery replaces the operator in the part preparation.

Furnace Brazing

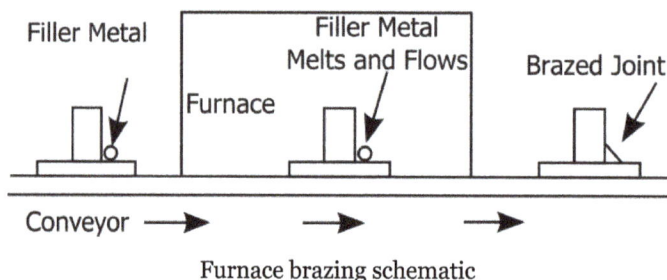

Furnace brazing schematic

Furnace brazing is a semi-automatic process used widely in industrial brazing operations due to its adaptability to mass production and use of unskilled labor. There are many advantages of furnace brazing over other heating methods that make it ideal for mass production. One main advantage is the ease with which it can produce large numbers of small parts that are easily jigged or self-locating. The process also offers the benefits of a controlled heat cycle (allowing use of parts that might distort under localized heating) and no need for post braze cleaning. Common atmospheres used include: inert, reducing or vacuum atmospheres all of which protect the part from oxidation. Some other advantages include: low unit cost when used in mass production, close temperature control, and the ability to braze multiple joints at once. Furnaces are typically heated using either electric, gas or oil depending on the type of furnace and application. However, some of the disadvantages of this method include: high capital equipment cost, more difficult design considerations and high power consumption.

There are four main types of furnaces used in brazing operations: batch type; continuous; retort with controlled atmosphere; and vacuum.

A *batch* type furnace has relatively low initial equipment costs, and can heat each part load separately. It can turned on and off at will, which reduces operating expenses when it's not in use. These furnaces are suited to medium to large volume production, and offer a large degree of flexibility in type of parts that can be brazed. Either controlled atmospheres or flux can be used to control oxidation and cleanliness of parts.

Continuous type furnaces are best suited to a steady flow of similar-sized parts through the furnace. These furnaces are often conveyor fed, moving parts through the hot zone at a controlled speed. It is common to use either controlled atmosphere or pre-applied flux in continuous furnaces. In particular, these furnaces offer the benefit of very low manual labor requirements and so are best suited to large scale production operations.

Retort-type furnaces differ from other batch-type furnaces in that they make use of a sealed lining called a "retort". The retort is generally sealed with either a gasket or is welded shut and filled completely with the desired atmosphere and then heated externally by conventional heating elements.

Due to the high temperatures involved, the retort is usually made of heat resistant alloys that resist oxidation. Retort furnaces are often either used in a batch or semi-continuous versions.

Vacuum furnaces is a relatively economical method of oxide prevention and is most often used to braze materials with very stable oxides (aluminum, titanium and zirconium) that cannot be brazed in atmosphere furnaces. Vacuum brazing is also used heavily with refractory materials and other exotic alloy combinations unsuited to atmosphere furnaces. Due to the absence of flux or a reducing atmosphere, the part cleanliness is critical when brazing in a vacuum. The three main types of vacuum furnace are: single-wall hot retort, double-walled hot retort, and cold-wall retort. Typical vacuum levels for brazing range from pressures of 1.3 to 0.13 pascals (10^{-2} to 10^{-3} Torr) to 0.00013 Pa (10^{-6} Torr) or lower. Vacuum furnaces are most commonly batch-type, and they are suited to medium and high production volumes.

Silver Brazing

Silver brazing, sometimes known as a *silver soldering* or *hard soldering*, is brazing using a silver alloy based filler. These silver alloys consist of many different percentages of silver and other metals, such as copper, zinc and cadmium.

Brazing is widely used in the tool industry to fasten 'hard metal' (carbide, ceramics, cermet, and similar) tips to tools such as saw blades. "Pretinning" is often done: the braze alloy is melted onto the hard metal tip, which is placed next to the steel and remelted. Pretinning gets around the problem that hard metals are hard to wet.

Brazed hard metal joints are typically two to seven mils thick. The braze alloy joins the materials and compensates for the difference in their expansion rates. It also provides a cushion between the hard carbide tip and the hard steel, which softens impact and prevents tip loss and damage—much as a vehicle's suspension helps prevent damage to the tires and the vehicle. Finally, the braze alloy joins the other two materials to create a composite structure, much as layers of wood and glue create plywood. The standard for braze joint strength in many industries is a joint that is stronger than either base material, so that when under stress, one or other of the base materials fails before the joint.

One special silver brazing method is called *pinbrazing* or *pin brazing*. It has been developed especially for connecting cables to railway track or for cathodic protection installations. The method uses a silver- and flux-containing brazing pin, which is melted in the eye of a cable lug. The equipment is normally powered from batteries.

Braze Welding

Braze welding is the use of a bronze or brass filler rod coated with flux to join steel workpieces. The equipment needed for braze welding is basically identical to the equipment used in brazing. Since braze welding usually requires more heat than brazing, acetylene or methylacetylene-propadiene (MAP) gas fuel is commonly used. The name comes from the fact that no capillary action is used.

Braze welding has many advantages over fusion welding. It allows the joining of dissimilar metals, minimization of heat distortion, and can reduce the need for extensive pre-heating. Additionally, since the metals joined are not melted in the process, the components retain their original shape;

edges and contours are not eroded or changed by the formation of a fillet. Another effect of braze welding is the elimination of stored-up stresses that are often present in fusion welding. This is extremely important in the repair of large castings. The disadvantages are the loss of strength when subjected to high temperatures and the inability to withstand high stresses.

Carbide, cermet and ceramic tips are plated and then joined to steel to make tipped band saws. The plating acts as a braze alloy.

Cast Iron "Welding"

The "welding" of cast iron is usually a brazing operation, with a filler rod made chiefly of nickel being used although true welding with cast iron rods is also available. Ductile cast iron pipe may be also "cadwelded," a process that connects joints by means of a small copper wire fused into the iron when previously ground down to the bare metal, parallel to the iron joints being formed as per hub pipe with neoprene gasket seals. The purpose behind this operation is to use electricity along the copper for keeping underground pipes warm in cold climates.

Vacuum Brazing

Vacuum brazing is a material joining technique that offers significant advantages: extremely clean, superior, flux-free braze joints of high integrity and strength. The process can be expensive because it must be performed inside a vacuum chamber vessel. Temperature uniformity is maintained on the work piece when heating in a vacuum, greatly reducing residual stresses due to slow heating and cooling cycles. This, in turn, can significantly improve the thermal and mechanical properties of the material, thus providing unique heat treatment capabilities. One such capability is heat-treating or age-hardening the workpiece while performing a metal-joining process, all in a single furnace thermal cycle.

Vacuum brazing is often conducted in a furnace; this means that several joints can be made at once because the whole workpiece reaches the brazing temperature. The heat is transferred using radiation, as many other methods cannot be used in a vacuum.

Dip Brazing

Dip brazing is especially suited for brazing aluminium because air is excluded, thus preventing the formation of oxides. The parts to be joined are fixtured and the brazing compound applied to the mating surfaces, typically in slurry form. Then the assemblies are dipped into a bath of molten salt (typically NaCl, KCl and other compounds), which functions as both heat transfer medium and flux.

Heating Methods

There are many heating methods available to accomplish brazing operations. The most important factor in choosing a heating method is achieving efficient transfer of heat throughout the joint and doing so within the heat capacity of the individual base metals used. The geometry of the braze joint is also a crucial factor to consider, as is the rate and volume of production required. The easiest way to categorize brazing methods is to group them by heating method. Here are some of the most common:

- Torch brazing

- Furnace brazing

- Induction brazing

- Dip brazing

- Resistance brazing

- Infrared brazing

- Blanket brazing

- Electron beam and laser brazing

- Braze welding

Advantages and Disadvantages

Brazing has many advantages over other metal-joining techniques, such as welding. Since brazing does not melt the base metal of the joint, it allows much tighter control over tolerances and produces a clean joint without the need for secondary finishing. Additionally, dissimilar metals and non-metals (i.e. metalized ceramics) can be brazed. In general, brazing also produces less thermal distortion than welding due to the uniform heating of a brazed piece. Complex and multi-part assemblies can be brazed cost-effectively. Welded joints must sometimes be ground flush, a costly secondary operation that brazing does not require because it produces a clean joint. Another advantage is that the brazing can be coated or clad for protective purposes. Finally, brazing is easily adapted to mass production and it is easy to automate because the individual process parameters are less sensitive to variation.

One of the main disadvantages is: the lack of joint strength as compared to a welded joint due to the softer filler metals used. The strength of the brazed joint is likely to be less than that of the base metal(s) but greater than the filler metal. Another disadvantage is that brazed joints can be damaged under high service temperatures. Brazed joints require a high degree of base-metal cleanliness when done in an industrial setting. Some brazing applications require the use of adequate fluxing agents to control cleanliness. The joint color is often different from that of the base metal, creating an aesthetic disadvantage.

Filler Metals

Some brazes come in the form of trifoils, laminated foils of a carrier metal clad with a layer of braze at each side. The center metal is often copper; its role is to act as a carrier for the alloy, to absorb mechanical stresses due to e.g. differential thermal expansion of dissimilar materials (e.g. a carbide tip and a steel holder), and to act as a diffusion barrier (e.g. to stop diffusion of aluminium from aluminium bronze to steel when brazing these two).

Braze Families

Brazing alloys form several distinct groups; the alloys in the same group have similar properties and uses.

- Pure metals

  Unalloyed. Often noble metals – silver, gold, palladium.

- Ag-Cu

  Silver-copper. Good melting properties. Silver enhances flow. Eutectic alloy used for furnace brazing. Copper-rich alloys prone to stress cracking by ammonia.

- Ag-Zn

  Silver-zinc. Similar to Cu-Zn, used in jewelry due to its high silver content so that the product is compliant with hallmarking. The color matches silver, and it is resistant to ammonia-containing silver-cleaning fluids.

- Cu-Zn (brass)

  Copper-zinc. General purpose, used for joining steel and cast iron. Corrosion resistance usually inadequate for copper, silicon bronze, copper-nickel, and stainless steel. Reasonably ductile. High vapor pressure due to volatile zinc, unsuitable for furnace brazing. Copper-rich alloys prone to stress cracking by ammonia.

- Ag-Cu-Zn

  Silver-copper-zinc. Lower melting point than Ag-Cu for same Ag content. Combines advantages of Ag-Cu and Cu-Zn. At above 40% Zn the ductility and strength drop, so only lower-zinc alloys of this type are used. At above 25% zinc less ductile copper-zinc and silver-zinc phases appear. Copper content above 60% yields reduced strength and liquidus above 900 °C. Silver content above 85% yields reduced strength, high liquidus and high cost. Copper-rich alloys prone to stress cracking by ammonia. Silver-rich brazes (above 67.5% Ag) are hallmarkable and used in jewellery; alloys with lower silver content are used for engineering purposes. Alloys with copper-zinc ratio of about 60:40 contain the same phases as brass and match its color; they are used for joining brass. Small amount of nickel improves strength and corrosion resistance and promotes wetting of carbides. Addition of manganese together with nickel increases fracture toughness. Addition of cadmium yields Ag-Cu-Zn-Cd alloys with improved fluidity and wetting and lower melting point; however cadmium is toxic. Addition of tin can play mostly the same role.

- Cu-P

  Copper-phosphorus. Widely used for copper and copper alloys. Does not require flux for copper. Can be also used with silver, tungsten, and molybdenum. Copper-rich alloys prone to stress cracking by ammonia.

- Ag-Cu-P

  Like Cu-P, with improved flow. Better for larger gaps. More ductile, better electrical conductivity. Copper-rich alloys prone to stress cracking by ammonia.

- Au-Ag

Gold-silver. Noble metals. Used in jewelry.

- Au-Cu

Gold-copper. Continuous series of solid solutions. Readily wet many metals, including refractory ones. Narrow melting ranges, good fluidity. Frequently used in jewellery. Alloys with 40–90% of gold harden on cooling but stay ductile. Nickel improves ductility. Silver lowers melting point but worsens corrosion resistance. To maintain corrosion resistance, gold must be kept above 60%. High-temperature strength and corrosion resistance can be improved by further alloying, e.g., with chromium, palladium, manganese, and molybdenum. Added vanadium allows wetting ceramics. Gold-copper has low vapor pressure.

- Au-Ni

Gold-Nickel. Continuous series of solid solutions. Wider melting range than Au-Cu alloys but better corrosion resistance and improved wetting. Frequently alloyed with other metals to reduce proportion of gold while maintaining properties. Copper may be added to lower gold proportion, chromium to compensate for loss of corrosion resistance, and boron for improving wetting impaired by the chromium. Generally no more than 35% Ni is used, as higher Ni/Au ratios have too wide melting range. Low vapor pressure.

- Au-Pd

Gold-Palladium. Improved corrosion resistance over Au-Cu and Au-Ni alloys. Used for joining superalloys and refractory metals for high-temperature applications, e.g. jet engines. Expensive. May be substituted for by cobalt-based brazes. Low vapor pressure.

- Pd

Palladium. Good high-temperature performance, high corrosion resistance (less than gold), high strength (more than gold). usually alloyed with nickel, copper, or silver. Forms solid solutions with most metals, does not form brittle intermetallics. Low vapor pressure.

- Ni

Nickel alloys, even more numerous than silver alloys. High strength. Lower cost than silver alloys. Good high-temperature performance, good corrosion resistance in moderately aggressive environments. Often used for stainless steels and heat-resistant alloys. Embrittled with sulfur and some lower-melting point metals, e.g. zinc. Boron, phosphorus, silicon and carbon lower melting point and rapidly diffuse to base metals. This allows diffusion brazing, and lets the joint be used above the brazing temperature. Borides and phosphides form brittle phases. Amorphous preforms can be made by rapid solidification.

- Co

Cobalt alloys. Good high-temperature corrosion resistance, possible alternative to Au-Pd brazes. Low workability at low temperatures, preforms prepared by rapid solidification.

- Al-Si

Aluminium-silicon. For brazing aluminium.

- Active alloys

Containing active metals, e.g., titanium or vanadium. Used for brazing non-metallic materials, e.g. graphite or ceramics.

Role of Elements

element	role	volatility	corrosion resistance	cost	incompatibility	description
Silver	structural, wetting	volatile		expensive		Enhances capillary flow, improves corrosion resistance of less-noble alloys, worsens corrosion resistance of gold and palladium. Relatively expensive. High vapor pressure, problematic in vacuum brazing. Wets copper. Does not wet nickel and iron. Reduces melting point of many alloys, including gold-copper.
Copper	structural				ammonia	Good mechanical properties. Often used with silver. Dissolves and wets nickel. Somewhat dissolves and wets iron. Copper-rich alloys sensitive to stress cracking in presence of ammonia.
Zinc	structural, melting, wetting	volatile	low	cheap	Ni	Lowers melting point. Often used with copper. Susceptible to corrosion. Improves wetting on ferrous metals and on nickel alloys. Compatible with aluminium. High vapor tension, produces somewhat toxic fumes, requires ventilation; highly volatile above 500 °C. At high temperatures may boil and create voids. Prone to selective leaching in some environments, which may cause joint failure. Traces of bismuth and beryllium together with tin or zinc in aluminium-based braze destabilize oxide film on aluminium, facilitating its wetting. High affinity to oxygen, promotes wetting of copper in air by reduction of the cuprous oxide surface film. Less such benefit in furnace brazing with controlled atmosphere. Embrittles nickel. High levels of zinc may result in a brittle alloy. Prone to interfacial corrosion in contact with stainless steel in wet and humid environments. Unsuitable for furnace brazing due to volatility.
Aluminium	structural, active				Fe	Usual base for brazing aluminium and its alloys. Embrittles ferrous alloys.
Gold	structural, wetting		excellent	very expensive		Excellent corrosion resistance. Very expensive. Wets most metals.
Palladium	structural		excellent	very expensive		Excellent corrosion resistance, though less than gold. Higher mechanical strength than gold. Good high-temperature strength. Very expensive, though less than gold. Makes the joint less prone to fail due to intergranular penetration when brazing alloys of nickel, molybdenum, or tungsten. Increases high-temperature strength of gold-based alloys. Improves high-temperature strength and corrosion resistance of gold-copper alloys. Forms solid solutions with most engineering metals, does not form brittle intermetallics. High oxidation resistance at high temperatures, especially Pd-Ni alloys.

Cadmium	structural, wetting, melting	volatile			toxic	Lowers melting point, improves fluidity. Toxic. Produces toxic fumes, requires ventilation. High affinity to oxygen, promotes wetting of copper in air by reduction of the cuprous oxide surface film. Less such benefit in furnace brazing with controlled atmosphere. Allows reducing silver content of Ag-Cu-Zn alloys. Replaced by tin in more modern alloys. In EU since December 2011 allowed only for aerospace and military use.
Lead	structural, melting					Lowers melting point. Toxic. Produces toxic fumes, requires ventilation.
Tin	structural, melting, wetting					Lowers melting point, improves fluidity. Broadens melting range. Can be used with copper, with which it forms bronze. Improves wetting of many difficult-to-wet metals, e.g. stainless steels and tungsten carbide. Traces of bismuth and beryllium together with tin or zinc in aluminium-based braze destabilize oxide film on aluminium, facilitating its wetting. Low solubility in zinc, which limits its content in zinc-bearing alloys.
Bismuth	trace additive					Lowers melting point. May disrupt surface oxides. Traces of bismuth and beryllium together with tin or zinc in aluminium-based braze destabilize oxide film on aluminium, facilitating its wetting.
Beryllium	trace additive				toxic	Traces of bismuth and beryllium together with tin or zinc in aluminium-based braze destabilize oxide film on aluminium, facilitating its wetting.
Nickel	structural, wetting		high		Zn, S	Strong, corrosion-resistant. Impedes flow of the melt. Addition to gold-copper alloys improves ductility and resistance to creep at high temperatures. Addition to silver allows wetting of silver-tungsten alloys and improves bond strength. Improves wetting of copper-based brazes. Improves ductility of gold-copper brazes. Improves mechanical properties and corrosion resistance of silver-copper-zinc brazes. Nickel content offsets brittleness induced by diffusion of aluminium when brazing aluminium-containing alloys, e.g. aluminium bronzes. In some alloys increases mechanical properties and corrosion resistance, by a combination of solid solution strengthening, grain refinement, and segregation on fillet surface and in grain boundaries, where it forms a corrosion-resistant layer. Extensive intersolubility with iron, chromium, manganese, and others; can severely erode such alloys. Embrittled by zinc, many other low melting point metals, and sulfur.
Chromium	structural		high			Corrosion-resistant. Increases high-temperature corrosion resistance and strength of gold-based alloys. Added to copper and nickel to increase corrosion resistance of them and their alloys. Wets oxides, carbides, and graphite; frequently a major alloy component for high-temperature brazing of such materials. Impairs wetting by gold-nickel alloys, which can be compensated for by addition of boron.

Manganese	structural	volatile	good	cheap		High vapor pressure, unsuitable for vacuum brazing. In gold-based alloys increases ductility. Increases corrosion resistance of copper and nickel alloys. Improves high-temperature strength and corrosion resistance of gold-copper alloys. Higher manganese content may aggravate tendency to liquation. Manganese in some alloys may tend to cause porosity in fillets. Tends to react with graphite molds and jigs. Oxidizes easily, requires flux. Lowers melting point of high-copper brazes. Improves mechanical properties and corrosion resistance of silver-copper-zinc brazes. Cheap, even less expensive than zinc. Part of the Cu-Zn-Mn system is brittle, some ratios can not be used. In some alloys increases mechanical properties and corrosion resistance, by a combination of solid solution strengthening, grain refinement, and segregation on fillet surface and in grain boundaries, where it forms a corrosion-resistant layer. Facilitates wetting of cast iron due to its ability to dissolve carbon. Improves conditions for brazing of carbides.
Molybdenum	structural		good			Increases high-temperature corrosion and strength of gold-based alloys. Increases ductility of gold-based alloys, promotes their wetting of refractory materials, namely carbides and graphite. When present in alloys being joined, may destabilize the surface oxide layer (by oxidizing and then volatilizing) and facilitate wetting.
Cobalt	structural		good			Good high-temperature properties and corrosion resistance. In nuclear applications can absorb neutrons and build up cobalt-60, a potent gamma radiation emitter.
Magnesium	volatile O_2 getter	volatile				Addition to aluminium makes the alloy suitable for vacuum brazing. Volatile, though less than zinc. Vaporization promotes wetting by removing oxides from the surface, vapors act as getter for oxygen in the furnace atmosphere.
Indium	melting, wetting			expensive		Lowers melting point. Improves wetting of ferrous alloys by copper-silver alloys. Suitable for joining parts that will be later coated by titanium nitride.
Carbon	melting					Lowers melting point. Can form carbides. Can diffuse to the base metal, resulting in higher remelt temperature, potentially allowing step-brazing with the same alloy. At above 0.1% worsens corrosion resistance of nickel alloys. Trace amounts present in stainless steel may facilitate reduction of surface chromium(III) oxide in vacuum and allow fluxless brazing. Diffusion away from the braze increases its remelt temperature; exploited in diffusion brazing.
Silicon	melting, wetting				Ni	Lowers melting point. Can form silicides. Improves wetting of copper-based brazes. Promotes flow. Causes intergranular embrittlement of nickel alloys. Rapidly diffuses into the base metals. Diffusion away from the braze increases its remelt temperature; exploited in diffusion brazing.
Germanium	structural, melting			expensive		Lowers melting point. Expensive. For special applications. May create brittle phases.

Boron	melting, wetting					Ni	Lowers melting point. Can form hard and brittle borides. Unsuitable for nuclear reactors, as boron is a potent neutron absorber and therefore acts as a neutron poison. Fast diffusion to the base metals. Can diffuse to the base metal, resulting in higher remelt temperature, potentially allowing step-brazing with the same alloy. Can erode some base materials or penetrate between grain boundaries of many heat-resistant structural alloys, degrading their mechanical properties. Causes intergranular embrittlement of nickel alloys. Improves wetting of/by some alloys, can be added to Au-Ni-Cr alloy to compensate for wetting loss by chromium addition. In low concentrations improves wetting and lowers melting point of nickel brazes. Rapidly diffuses to base materials, may lower their melting point; especially a concern when brazing thin materials. Diffusion away from the braze increases its remelt temperature; exploited in diffusion brazing.
Mischmetal	trace additive						in amount of about 0.08%, can be used to substitute boron where boron would have detrimental effects.
Cerium	trace additive						in trace quantities, improves fluidity of brazes. Particularly useful for alloys of four or more components, where the other additives compromise flow and spreading.
Strontium	trace additive						in trace quantities, refines the grain structure of aluminium-based alloys.
Phosphorus	deoxidizer					H_2S, SO_2, Ni, Fe, Co	Lowers melting point. Deoxidizer, decomposes copper oxide; phosphorus-bearing alloys can be used on copper without flux. Does not decompose zinc oxide, so flux is needed for brass. Forms brittle phosphides with some metals, e.g. nickel (Ni_3P) and iron, phosphorus alloys unsuitable for brazing alloys bearing iron, nickel or cobalt in amount above 3%. The phosphides segregate at grain boundaries and cause intergranular embrittlement. (Sometimes the brittle joint is actually desired, though. Fragmentation grenades can be brazed with phosphorus bearing alloy to produce joints that shatter easily at detonation.) Avoid in environments with presence of sulfur dioxide (e.g. paper mills) and hydrogen sulfide (e.g. sewers, or close to volcanoes); the phosphorus-rich phase rapidly corrodes in presence of sulfur and the joint fails. Phosphorus can be also present as an impurity introduced from e.g. electroplating baths. In low concentrations improves wetting and lowers melting point of nickel brazes. Diffusion away from the braze increases its remelt temperature; exploited in diffusion brazing.
Lithium	deoxidizer						Deoxidizer. Eliminates the need for flux with some materials. Lithium oxide formed by reaction with the surface oxides is easily displaced by molten braze alloy.
Titanium	structural, active						Most commonly used active metal. Few percents added to Ag-Cu alloys facilitate wetting of ceramics, e.g. silicon nitride. Most metals, except few (namely silver, copper and gold), form brittle phases with titanium. When brazing ceramics, like other active metals, titanium reacts with them and forms a complex layer on their surface, which in turn is wettable by the silver-copper braze. Wets oxides, carbides, and graphite; frequently a major alloy component for high-temperature brazing of such materials.

Zirconium	structural, active					Wets oxides, carbides, and graphite; frequently a major alloy component for high-temperature brazing of such materials.
Hafnium	active					
Vanadium	structural, active					Promotes wetting of alumina ceramics by gold-based alloys.
Sulfur	impurity					Compromises integrity of nickel alloys. Can enter the joints from residues of lubricants, grease or paint. Forms brittle nickel sulfide (Ni_3S_2) that segregates at grain boundaries and cause intergranular failure.

Some additives and impurities act at very low levels. Both positive and negative effects can be observed. Strontium at levels of 0.01% refines grain structure of aluminium. Beryllium and bismuth at similar levels help disrupt the passivation layer of aluminium oxide and promote wetting. Carbon at 0.1% impairs corrosion resistance of nickel alloys. Aluminium can embrittle mild steel at 0.001%, phosphorus at 0.01%.

In some cases, especially for vacuum brazing, high-purity metals and alloys are used. 99.99% and 99.999% purity levels are available commercially.

Care must be taken to not introduce deleterious impurities from joint contamination or by dissolution of the base metals during brazing.

Melting Behavior

Alloys with larger span of solidus/liquidus temperatures tend to melt through a "mushy" state, during which the alloy is a mixture of solid and liquid material. Some alloys show tendency to liquation, separation of the liquid from the solid portion; for these the heating through the melting range must be sufficiently fast to avoid this effect. Some alloys show extended plastic range, when only a small portion of the alloy is liquid and most of the material melts at the upper temperature range; these are suitable for bridging large gaps and for forming fillets. Highly fluid alloys are suitable for penetrating deep into narrow gaps and for brazing tight joints with narrow tolerances but are not suitable for filling larger gaps. Alloys with wider melting range are less sensitive to non-uniform clearances.

When the brazing temperature is suitably high, brazing and heat treatment can be done in a single operation simultaneously.

Eutectic alloys melt at single temperature, without mushy region. Eutectic alloys have superior spreading; non-eutectics in the mushy region have high viscosity and at the same time attack the base metal, with correspondingly lower spreading force. Fine grain size gives eutectics both increased strength and increased ductility. Highly accurate melting temperature lets joining process be performed only slightly above the alloy's melting point. On solidifying, there is no mushy state where the alloy appears solid but is not yet; the chance of disturbing the joint by manipulation in such state is reduced (assuming the alloy did not significantly change its properties by dissolving the base metal). Eutectic behavior is especially beneficial for solders.

Metals with fine grain structure before melting provide superior wetting to metals with large grains. Alloying additives (e.g. strontium to aluminium) can be added to refine grain structure, and the preforms or foils can be prepared by rapid quenching. Very rapid quenching may provide amorphous metal structure, which possess further advantages.

Interaction with Base Metals

Brazing at the Gary Tubular Steel Plant, 1943

For successful wetting, the base metal must be at least partially soluble in at least one component of the brazing alloy. The molten alloy therefore tends to attack the base metal and dissolve it, slightly changing its composition in the process. The composition change is reflected in the change of the alloy's melting point and the corresponding change of fluidity. For example, some alloys dissolve both silver and copper; dissolved silver lowers their melting point and increases fluidity, copper has the opposite effect.

The melting point change can be exploited. As the remelt temperature can be increased by enriching the alloy with dissolved base metal, step brazing using the same braze can be possible.

Alloys that do not significantly attack the base metals are more suitable for brazing thin sections.

Nonhomogenous microstructure of the braze may cause non-uniform melting and localized erosions of the base metal.

Wetting of base metals can be improved by adding a suitable metal to the alloy. Tin facilitates wetting of iron, nickel, and many other alloys. Copper wets ferrous metals that silver does not attack, copper-silver alloys can therefore braze steels silver alone won't wet. Zinc improves wetting of ferrous metals, indium as well. Aluminium improves wetting of aluminium alloys. For wetting of ceramics, reactive metals capable of forming chemical compounds with the ceramic (e.g. titanium, vanadium, zirconium...) can be added to the braze.

Dissolution of base metals can cause detrimental changes in the brazing alloy. For example, aluminium dissolved from aluminium bronzes can embrittle the braze; addition of nickel to the braze can offset this.

The effect works both ways; there can be detrimental interactions between the braze alloy and the base metal. Presence of phosphorus in the braze alloy leads to formation of brittle phosphides of iron and nickel, phosphorus-containing alloys are therefore unsuitable for brazing nickel and ferrous alloys. Boron tends to diffuse into the base metals, especially along the grain boundaries, and may form brittle borides. Carbon can negatively influence some steels.

Care must be taken to avoid galvanic corrosion between the braze and the base metal, and especially between dissimilar base metals being brazed together. Formation of brittle intermetallic compounds on the alloy interface can cause joint failure. This is discussed more in-depth with solders.

The potentially detrimental phases may be distributed evenly through the volume of the alloy, or be concentrated on the braze-base interface. A thick layer of interfacial intermetallics is usually considered detrimental due to its commonly low fracture toughness and other sub-par mechanical properties. In some situations, e.g. die attaching, it however does not matter much as silicon chips are not typically subjected to mechanical abuse.

On wetting, brazes may liberate elements from the base metal. For example, aluminium-silicon braze wets silicon nitride, dissociates the surface so it can react with silicon, and liberates nitrogen, which may create voids along the joint interface and lower its strength. Titanium-containing nickel-gold braze wets silicon nitride and reacts with its surface, forming titanium nitride and liberating silicon; silicon then forms brittle nickel silicides and eutectic gold-silicon phase; the resulting joint is weak and melts at much lower temperature than may be expected.

Metals may diffuse from one base alloy to the other one, causing embrittlement or corrosion. An example is diffusion of aluminium from aluminium bronze to a ferrous alloy when joining these. A diffusion barrier, e.g. a copper layer (e.g. in a trimet strip), can be used.

A sacrificial layer of a noble metal can be used on the base metal as an oxygen barrier, preventing formation of oxides and facilitating fluxless brazing. During brazing, the noble metal layer dissolves in the filler metal. Copper or nickel plating of stainless steels performs the same function.

In brazing copper, a reducing atmosphere (or even a reducing flame) may react with the oxygen residues in the metal, which are present as cuprous oxide inclusions, and cause hydrogen embrittlement. The hydrogen present in the flame or atmosphere at high temperature reacts with the oxide, yielding metallic copper and water vapour, steam. The steam bubbles exert high pressure in the metal structure, leading to cracks and joint porosity. Oxygen-free copper is not sensitive to this effect, however the most readily available grades, e.g. electrolytic copper or high-conductivity copper, are. The embrittled joint may then fail catastrophically without any previous sign of deformation or deterioration.

Preform

A brazing preform is a high quality, precision metal stamping used for a variety of joining applications in manufacturing electronic devices and systems. Typical brazing preform uses include attaching electronic circuitry, packaging electronic devices, providing good thermal and electrical conductivity, and providing an interface for electronic connections. Square, rectangular and disc shaped brazing preforms are commonly used to attach electronic components containing silicon dies to a substrate such as a printed circuit board.

Rectangular frame shaped preforms are often required for the construction of electronic packages while washer shaped brazing preforms are typically utilized to attach lead wires and hermetic feedthroughs to electronic circuits and packages. Some preforms are also used in diodes, rectifiers, optoelectronic devices and components packaging.

Induction Brazing

Induction brazing is a process in which two or more materials are joined together by a filler metal that has a lower melting point than the base materials using induction heating. In induction heating, usually ferrous materials are heated rapidly from the electromagnetic field that is created by the alternating current from an induction coil.

Materials and Applications

"Induction brazing is suitable for many metallic materials, with magnetic materials being heated more readily. Where ceramic materials are involved, heating will most likely occur by conduction from surrounding metallic parts, or the use of a susceptor" (Sue Dunkerton, 1).

According to Ambrell Group Application Labs talking about filler metals: Silver is frequently used for induction brazing because of its low melting point. Silver-copper eutectic brazes have melting temperatures between 1100°F and 1650°F. Aluminum braze, the least common, has a melting temperature of 1050°F to 1140°F. Copper braze, the least expensive, has a melting temperature of 1300°F to 2150°F. (p1)

The filler can be manually applied but because of the more common semiautomatic production a preloaded joint is more commonly used to speed the operation and help to keep a more uniform bond.

Benefits

There are specific reasons to use induction heating for industrial brazing. These include selective heating, better joint quality, reduced oxidation and acid cleaning, faster heating cycles, more consistent results and suitability for large volume production.

Selective Heating

Induction heating can be targeted to provide heat to very small areas within tight production tolerances. Only those areas of the part within close proximity to the joint are heated; the rest of the part is not affected. Since there is no direct contact with the part, there is no opportunity for breakage. The life of the fixturing is substantially increased because problems due to repeated exposure to heat (such as distortion and metal fatigue) are eliminated. This advantage becomes particularly important with high-temperature brazing processes.

With efficient coil design, careful fixturing and consistent part placement, it is possible to simultaneously provide heat in different areas of the same part

Better Quality Joints

Induction heating produces clean, leak proof joints by preventing the filler from flowing in areas that it shouldn't flow. This ability to create clean and controllable joints is one of the reasons that induction brazing is being used extensively for high-precision, high-reliability applications.

Reduced Oxidation and Cleaning

Flame heating in a normal atmosphere causes oxidation, scaling and carbon build up on the parts. To clean the parts, applications of joint-weakening flux and expensive acid cleaning baths have traditionally been required. Batch vacuum furnaces solve these problems, but have significant limitations of their own because of their large size, poor efficiency and lack of quality control. Brazing with induction reduces both oxidation and costly cleaning requirements, especially when a rapid cool-down cycle is used.

Fast Heating Cycles

Because the induction heating cycle is very short in comparison to flame brazing, more parts can be processed in the same amount of time, and less heat is released to the surrounding environment. "An induction brazing system quickly delivers highly localized heat to minimize part warpage and distortion. Brazing in a controlled vacuum or in an inert protective atmosphere can significantly improve overall part quality and eliminate costly part cleaning procedures" (Induction Atmospheres, 1).

Consistent Results

Induction brazing is a very repeatable process because variables such as time, temperature, alloy, fixturing, and part positioning are very controllable. The internal power supply of the RF power supply can be used to control cycle time, and temperature control can be accomplished with pyrometers, visual temperature sensors or thermocouples.

For processes, which involve medium to high production runs of the same parts, an automated part handling system is often utilized to further improve consistency and maximize productivity. For the most part, induction brazing and soldering is done in an open-air environment but it can also be done in a controlled atmosphere when necessary to keep the parts completely clean and free of oxidation. Induction brazing generally works best with two pieces of similar metal. Dissimilar metals can also be joined by induction heating but they require special attention and techniques. This is due to differences in the materials' resistivity, relative magnetic permeability and coefficients of thermal expansion. (p1)

General Temperatures and Times

According to Turnkey Induction Heating Solutions:

Process	Time	Temperature (°F)
Brazing Stainless Steel Tubes	20 seconds	1330°F
Brazing Stainless Steel Orthodontic Parts	1 second	1300°F
Brazing Hydraulic Hose Assemblies	7 seconds	2200°F
Brazing Metering Plates to Turbine Blades With Nickel	5 minutes	2000°F

Brazing Copper Tube Assemblies	45 seconds	1450°F
Brazing Stainless Steel to Brass	7 seconds	1325°F
Brazing Stainless Steel to Titanium	80 seconds	2000°F
Brazing Stainless Steel Dental Tools	10 seconds	1400°F(p1)

Soldering

Desoldering a contact from a wire

Soldering is a process in which two or more items (usually metal) are joined together by melting and putting a filler metal (solder) into the joint, the filler metal having a lower melting point than the adjoining metal. Soldering differs from welding in that soldering does not involve melting the work pieces. In brazing, the filler metal melts at a higher temperature, but the work piece metal does not melt. In the past, nearly all solders contained lead, but environmental and health concerns have increasingly dictated use of lead-free alloys for electronics and plumbing purposes.

Origins

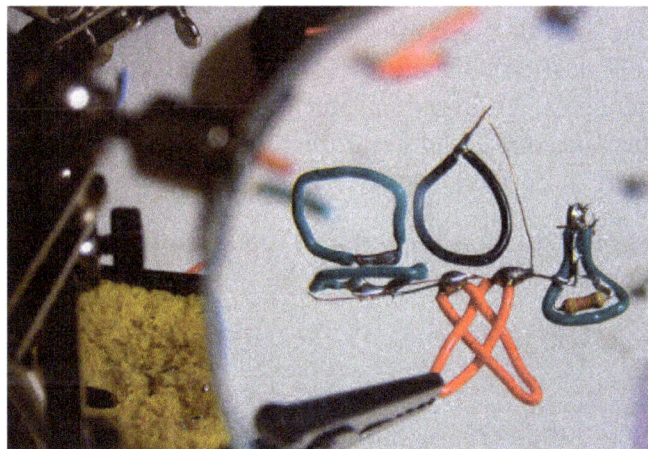

Small figurine being created by soldering

There is evidence that soldering was employed as early as 5000 years ago in Mesopotamia. Soldering and brazing are thought to have originated very early in the history of metal-working, probably before 4000 BC. Sumerian swords from ~3000 BC were assembled using hard soldering.

Soldering was historically used to make jewelry items, cooking ware and tools, as well as other uses such as in assembling stained glass.

Applications

Soldering is used in plumbing, electronics, and metalwork from flashing to jewelry.

Soldering provides reasonably permanent but reversible connections between copper pipes in plumbing systems as well as joints in sheet metal objects such as food cans, roof flashing, rain gutters and automobile radiators.

Jewelry components, machine tools and some refrigeration and plumbing components are often assembled and repaired by the higher temperature silver soldering process. Small mechanical parts are often soldered or brazed as well. Soldering is also used to join lead came and copper foil in stained glass work.

Electronic soldering connects electrical wiring and electronic components to printed circuit boards (PCBs).

Solders

Soldering filler materials are available in many different alloys for differing applications. In electronics assembly, the eutectic alloy of 63% tin and 37% lead (or 60/40, which is almost identical in melting point) has been the alloy of choice. Other alloys are used for plumbing, mechanical assembly, and other applications. Some examples of soft-solder are tin-lead for general purposes, tin-zinc for joining aluminium, lead-silver for strength at higher than room temperature, cadmium-silver for strength at high temperatures, zinc-aluminium for aluminium and corrosion resistance, and tin-silver and tin-bismuth for electronics.

A eutectic formulation has advantages when applied to soldering: the liquidus and solidus temperatures are the same, so there is no plastic phase, and it has the lowest possible melting point. Having the lowest possible melting point minimizes heat stress on electronic components during soldering. And, having no plastic phase allows for quicker wetting as the solder heats up, and quicker setup as the solder cools. A non-eutectic formulation must remain still as the temperature drops through the liquidus and solidus temperatures. Any movement during the plastic phase may result in cracks, resulting in an unreliable joint.

Common solder formulations based on tin and lead are listed below. The fraction represent percentage of tin first, then lead, totaling 100%:

- 63/37: melts at 183 °C (361 °F) (eutectic: the only mixture that melts at a *point*, instead of over a range)

- 60/40: melts between 183–190 °C (361–374 °F)

- 50/50: melts between 183–215 °C (361–419 °F)

For environmental reasons (and the introduction of regulations such as the European RoHS (Restriction of Hazardous Substances Directive)), lead-free solders are becoming more widely used. They are also suggested anywhere young children may come into contact with (since young children are likely to place things into their mouths), or for outdoor use where rain and other precipitation may wash the lead into the groundwater. Unfortunately, most lead-free solders are not eutectic formulations, melting at around 250 °C (482 °F), making it more difficult to create reliable joints with them.

Other common solders include low-temperature formulations (often containing bismuth), which are often used to join previously-soldered assemblies without un-soldering earlier connections, and high-temperature formulations (usually containing silver) which are used for high-temperature operation or for first assembly of items which must not become unsoldered during subsequent operations. Alloying silver with other metals changes the melting point, adhesion and wetting characteristics, and tensile strength. Of all the brazing alloys, silver solders have the greatest strength and the broadest applications. Specialty alloys are available with properties such as higher strength, the ability to solder aluminum, better electrical conductivity, and higher corrosion resistance.

Flux

The purpose of flux is to facilitate the soldering process. One of the obstacles to a successful solder joint is an impurity at the site of the joint, for example, dirt, oil or oxidation. The impurities can be removed by mechanical cleaning or by chemical means, but the elevated temperatures required to melt the filler metal (the solder) encourages the work piece (and the solder) to re-oxidize. This effect is accelerated as the soldering temperatures increase and can completely prevent the solder from joining to the workpiece. One of the earliest forms of flux was charcoal, which acts as a reducing agent and helps prevent oxidation during the soldering process. Some fluxes go beyond the simple prevention of oxidation and also provide some form of chemical cleaning (corrosion).

For many years, the most common type of flux used in electronics (soft soldering) was rosin-based, using the rosin from selected pine trees. It was ideal in that it was non-corrosive and non-conductive at normal temperatures but became mildly reactive (corrosive) at the elevated soldering temperatures. Plumbing and automotive applications, among others, typically use an acid-based (hydrochloric acid) flux which provides cleaning of the joint. These fluxes cannot be used in electronics because they are conductive and because they will eventually dissolve the small diameter wires. Many fluxes also act as a wetting agent in the soldering process, reducing the surface tension of the molten solder and causing it to flow and wet the workpieces more easily.

Fluxes for soft solder are currently available in three basic formulations:

1. Water-soluble fluxes - higher activity fluxes designed to be removed with water after soldering (no VOCs required for removal).

2. No-clean fluxes - mild enough to not "require" removal due to their non-conductive and non-corrosive residue. These fluxes are called "no-clean" because the residue left after the solder operation is non-conductive and won't cause electrical shorts; nevertheless they leave a plainly visible white residue that resembles diluted bird-droppings. No-clean flux

residue is acceptable on all 3 classes of PCBs as defined by IPC-610 provided it does not inhibit visual inspection, access to test points, or have a wet, tacky or excessive residue that may spread onto other areas. Connector mating surfaces must also be free of flux residue. Fingerprints in no-clean residue are a class 3 defect

3. Traditional rosin fluxes - available in non-activated (R), mildly activated (RMA) and activated (RA) formulations. RA and RMA fluxes contain rosin combined with an activating agent, typically an acid, which increases the wettability of metals to which it is applied by removing existing oxides. The residue resulting from the use of RA flux is corrosive and must be cleaned. RMA flux is formulated to result in a residue which is not significantly corrosive, with cleaning being preferred but optional.

Flux performance needs to be carefully evaluated; a very mild 'no-clean' flux might be perfectly acceptable for production equipment, but not give adequate performance for a poorly controlled hand-soldering operation.

Processes

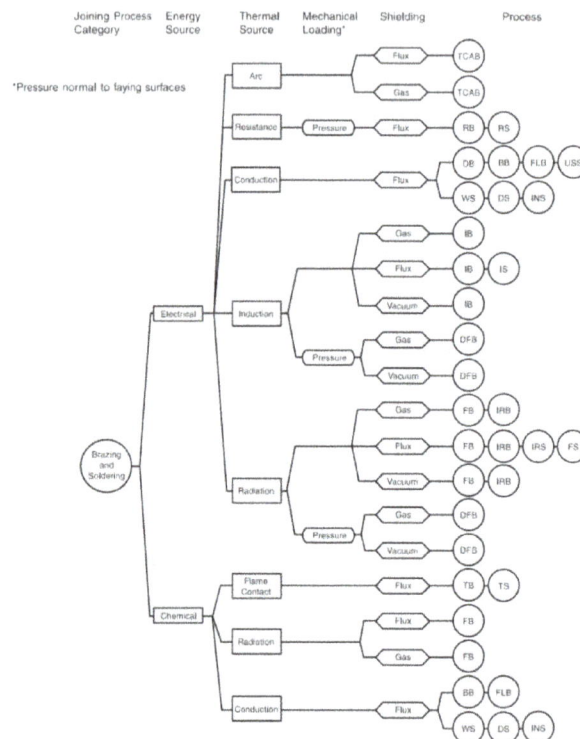

Brazing and soldering processes classification chart

There are three forms of soldering, each requiring progressively higher temperatures and producing an increasingly stronger joint strength:

1. soft soldering, which originally used a tin-lead alloy as the filler metal

2. silver soldering, which uses an alloy containing silver

3. brazing which uses a brass alloy for the filler

The alloy of the filler metal for each type of soldering can be adjusted to modify the melting temperature of the filler. Soldering differs from gluing significantly in that the filler metals alloy with the workpiece at the junction to form a gas- and liquid-tight bond.

Soft soldering is characterized by having a melting point of the filler metal below approximately 400 °C (752 °F), whereas silver soldering and brazing use higher temperatures, typically requiring a flame or carbon arc torch to achieve the melting of the filler. Soft solder filler metals are typically alloys (often containing lead) that have liquidus temperatures below 350 °C.

In this soldering process, heat is applied to the parts to be joined, causing the solder to melt and to bond to the workpieces in an alloying process called wetting. In stranded wire, the solder is drawn up into the wire by capillary action in a process called 'wicking'. Capillary action also takes place when the workpieces are very close together or touching. The joint's tensile strength is dependent on the filler metal used. Soldering produces electrically-conductive, water- and gas-tight joints.

Each type of solder offers advantages and disadvantages. Soft solder is so called because of the soft lead that is its primary ingredient. Soft soldering uses the lowest temperatures but does not make a strong joint and is unsuitable for mechanical load-bearing applications. It is also unsuitable for high-temperature applications as it softens and melts. Silver soldering, as used by jewelers, machinists and in some plumbing applications, requires the use of a torch or other high-temperature source, and is much stronger than soft soldering. Brazing provides the strongest joint but also requires the hottest temperatures to melt the filler metal, requiring a torch or other high temperature source and darkened goggles to protect the eyes from the bright light produced by the white-hot work. It is often used to repair cast-iron objects, wrought-iron furniture, etc.

Soldering operations can be performed with hand tools, one joint at a time, or *en masse* on a production line. Hand soldering is typically performed with a soldering iron, soldering gun, or a torch, or occasionally a hot-air pencil. Sheetmetal work was traditionally done with "soldering coppers" directly heated by a flame, with sufficient stored heat in the mass of the soldering copper to complete a joint; torches or electrically-heated soldering irons are more convenient. All soldered joints require the same elements of cleaning of the metal parts to be joined, fitting up the joint, heating the parts, applying flux, applying the filler, removing heat and holding the assembly still until the filler metal has completely solidified. Depending on the nature of flux material used, cleaning of the joints may be required after they have cooled.

Each alloy has characteristics that work best for certain applications, notably strength and conductivity, and each type of solder and alloy has different melting temperatures. The term *silver solder* likewise denotes the type of solder that is used. Some soft solders are "silver-bearing" alloys used to solder silver-plated items. Lead-based solders should not be used on precious metals because the lead dissolves the metal and disfigures it.

Soldering and Brazing

The distinction between soldering and brazing is based on the melting temperature of the filler alloy. A temperature of 450 °C is usually used as a practical delineating point between soldering and brazing . Soft soldering can be done with a heated iron whereas the other methods require a higher temperature torch or furnace to melt the filler metal.

Different equipment is usually required since a soldering iron cannot achieve high enough temperatures for hard soldering or brazing. Brazing filler metal is stronger than silver solder, which is stronger than lead-based soft solder. Brazing solders are formulated primarily for strength, silver solder is used by jewelers to protect the precious metal and by machinists and refrigeration technicians for its tensile strength but lower melting temperature than brazing, and the primary benefit of soft solder is the low temperature used (to prevent heat damage to electronic components and insulation).

Since the joint is produced using a metal with a lower melting temperature than the workpiece, the joint will weaken as the ambient temperature approaches the melting point of the filler metal. For that reason, the higher temperature processes produce joints which are effective at higher temperatures. Brazed connections can be as strong or nearly as strong as the parts they connect, even at elevated temperatures.

Silver Soldering

"Hard soldering" or "silver soldering" is used to join precious and semi-precious metals such as gold, silver, brass, and copper. The solder is usually referred to as easy, medium, or hard. This refers to its melting temperature, not the strength of the joint. Extra-easy solder contains 56% silver and has a melting point of 1,145 °F (618 °C). Extra-hard solder has 80% silver and melts at 1,370 °F (740 °C). If multiple joints are needed, then the jeweler will start with hard or extra-hard solder and switch to lower-temperature solders for later joints.

Silver solder is absorbed by the surrounding metal, resulting in a joint that is actually stronger than the metal being joined. The metal being joined must be perfectly flush, as silver solder cannot normally be used as a filler and any gaps will remain.

Another difference between brazing and soldering is how the solder is applied. In brazing, one generally uses rods that are touched to the joint while being heated. With silver soldering, small pieces of solder wire are placed onto the metal prior to heating. A flux, often made of boric acid and denatured alcohol, is used to keep the metal and solder clean and to prevent the solder from moving before it melts.

When silver solder melts, it tends to flow towards the area of greatest heat. Jewelers can somewhat control the direction the solder moves by leading it with a torch; it will even run straight up along a seam.

Induction Soldering

Induction soldering uses induction heating by high-frequency AC current in a surrounding copper coil. This induces currents in the part being soldered, which generates heat because of the higher resistance of a joint versus its surrounding metal (resistive heating). These copper coils can be shaped to fit the joint more precisely. A filler metal (solder) is placed between the facing surfaces, and this solder melts at a fairly low temperature. Fluxes are commonly used in induction soldering. This technique is particularly suited to continuously soldering, in which case these coils wrap around a cylinder or a pipe that needs to be soldered.

Some metals are easier to solder than others. Copper, silver, and gold are easy. Iron, mild steel and nickel are next in difficulty. Because of their thin, strong oxide films, stainless steel and aluminium are even more difficult to solder. Titanium, magnesium, cast irons, some high-carbon steels, ceramics, and graphite can be soldered but it involves a process similar to joining carbides: they are first plated with a suitable metallic element that induces interfacial bonding.

Electronic Components (PCBs)

Soldering of an SMD capacitor

A tube of multicore electronics solder used for manual soldering

An improperly soldered 'cold' joint

Broken solder joints on a circuit board

Currently, mass-production printed circuit boards (PCBs) are mostly wave soldered or reflow soldered, though hand soldering of production electronics is also still standard practice.

In wave soldering, parts are temporarily kept in place with small dabs of adhesive, then the assembly is passed over flowing solder in a bulk container. This solder is shaken into waves so the whole

PCB is not submerged in solder, but rather touched by these waves. The end result is that solder stays on pins and pads, but not on the PCB itself.

Reflow soldering is a process in which a solder paste (a mixture of prealloyed solder powder and a flux-vehicle that has a peanut butter-like consistency) is used to stick the components to their attachment pads, after which the assembly is heated by an infrared lamp, a hot air pencil, or, more commonly, by passing it through a carefully controlled oven.

Since different components can be best assembled by different techniques, it is common to use two or more processes for a given PCB. For example, surface mounted parts may be reflow soldered first, with a wave soldering process for the through-hole mounted components coming next, and bulkier parts hand-soldered last.

For hand soldering, the heat source tool should be selected to provide adequate heat for the size of joint to be completed. A 100 watt soldering iron may provide too much heat for printed circuit boards, while a 25 watt iron will not provide enough heat for large electrical connectors, joining copper roof flashing, or large stained-glass lead came. Using a tool with too high a temperature can damage sensitive components, but protracted heating by a tool that is too cool or under powered can also cause extensive heat damage.

Hand-soldering techniques require a great deal of skill to use on what is known as fine-pitch soldering of chip packages. In particular ball grid array (BGA) devices are notoriously difficult, if not impossible, to rework by hand.

For attachment of electronic components to a PCB, proper selection and use of flux helps prevent oxidation during soldering, which is essential for good wetting and heat transfer. The soldering iron tip must be clean and pre-tinned with solder to ensure rapid heat transfer. Components which dissipate large amounts of heat during operation are sometimes elevated above the PCB to avoid PCB overheating. After inserting a through-hole mounted component, the excess lead is cut off, leaving a length of about the radius of the pad. Plastic or metal mounting clips or holders may be used with large devices to aid heat dissipation and reduce joint stresses.

A heat sink may be used on the leads of heat sensitive components to reduce heat transfer to the component. This is especially applicable to germanium parts. (Note the heat sink will mean the use of more heat to complete the joint.) If all metal surfaces are not properly fluxed and brought above the melting temperature of the solder in use, the result will be an unreliable "cold solder joint".

To simplify soldering, beginners are usually advised to apply the soldering iron and the solder separately to the joint, rather than the solder being applied direct to the iron. When sufficient solder is applied, the solder wire is removed. When the surfaces are adequately heated, the solder will flow around the joint. The iron is then removed from the joint.

Since non-eutectic solder alloys have a small plastic range, the joint must not be moved until the solder has cooled down through both the liquidus and solidus temperatures. When visually inspected, a good solder joint will appear smooth and shiny, with the outline of the soldered wire clearly visible. A matte gray surface is a good indicator of a joint that was moved during soldering.

Other solder defects can be detected visually as well. Too little solder will result in a dry and unreliable joint; too much solder (the familiar 'solder blob' to beginners) is not necessarily unsound, but tends to mean poor wetting. With some fluxes, flux residue remaining on the joint may need to be removed, using water, alcohol or other solvents compatible with the parts in question.

Excess solder and unconsumed flux and residue is sometimes wiped from the soldering iron tip between joints. The tip of the iron is kept wetted with solder ("tinned") when hot to assist soldering, and when hot and cold to minimize oxidation and corrosion of the tip itself.

Environmental legislation in many countries, and the whole of the European Community area, has led to a change in formulation of both solders and fluxes. Water-soluble non-rosin-based fluxes have been increasingly used since the 1980s so that soldered boards can be cleaned with water or water-based cleaners. This eliminates hazardous solvents from the production environment, and from factory effluents.

Hot-bar Reflow

Hot-bar reflow is a selective soldering process where two pre-fluxed, solder coated parts are heated with heating element (called a thermode) to a sufficient temperature to melt the solder.

Pressure is applied through the whole process (usually 15 s) to ensure that components stay in place during cooling. The heating element is heated and cooled for each connection. Up to 4000 W can be used in the heating element allowing fast soldering, good results with connections requiring high energy.

Laser

Laser soldering is a technique where a ~30-50 W laser is used to melt and solder an electrical connection joint. Diode laser systems based on semiconductor junctions are used for this purpose. Suzanne Jenniches patented laser soldering in 1980.

Wavelengths are typically 808 nm through 980 nm. The beam is delivered via an optical fiber to the workpiece, with fiber diameters 800 μm and smaller. Since the beam out of the end of the fiber diverges rapidly, lenses are used to create a suitable spot size on the workpiece at a suitable working distance. A wire feeder is used to supply solder.

Both lead-tin and silver-tin material can be soldered. Process recipes will differ depending on the alloy composition. For soldering 44-pin chip carriers to a board using soldering preforms, power levels were on the order of 10 Watts and solder times approximately 1 second. Low power levels can lead to incomplete wetting and the formation of voids, both of which can weaken the joint.

Fiber Focus Infrared Soldering

Fiber focus infrared soldering is technique where many infrared sources are led through fibers, then focused onto a single spot at which the connection is soldered.

Pipe Soldering

Copper pipe, or 'tube', is commonly joined by soldering. When applied in a plumbing trade context in the United States, soldering is often referred to as *sweating*, and a tubing connection so made is referred to as a *sweated joint*.

Soldered copper pipes

Solder

Lead-free solder

Copper tubing conducts heat away much faster than a conventional hand-held soldering iron or gun can provide, so a propane torch is most commonly used to deliver the necessary power; for large tubing sizes and fittings a MAPP-fueled, acetylene-fueled, or propylene-fueled torch is used with atmospheric air as the oxidizer; MAPP/oxygen or acetylene/oxygen are rarely used because the flame temperature is much higher than the melting point of copper. Too much heat destroys the temper of hard-tempered copper tubing, and can burn the flux out of a joint before the solder is added, resulting in a faulty joint. For larger tubing sizes, a torch fitted with various sizes of inter-changeable *swirl tips* is employed to deliver the needed heating power. Most experienced plumb-ers seldom use propane fuel. In the hands of a skilled tradesman, the hotter flame of acetylene, MAPP, or propylene allows more joints to be completed per hour.

However, it is possible to use an electrical tool to solder joints in copper pipe sized from 8mm to 22mm. For example, the Antex Pipemaster is recommended for use in tight spaces, when open flames are hazardous, or by do-it-yourself users. The pliers-like tool uses heated fitted jaws that completely encircle the pipe, allowing a joint to be melted in as little as 10 seconds.

Solder fittings, also known as *capillary fittings*, are short sections of smooth pipe designed to slide over the outside of the mating tube, are usually used for copper joints. Commonly used fittings include for straight connectors, reducers, bends, and tees. There are two types of solder fittings: *end feed fittings* which contain no solder, and *solder ring fittings* (also known as Yorkshire fittings), in which there is a ring of solder in a small circular recess inside the fitting.

As with all solder joints, all parts to be joined must be clean and oxide free. Internal and external wire brushes are available for the common pipe and fitting sizes; emery cloth and wire-wool are frequently used as well, although metal wool products are discouraged, as they can contain oil, which would contaminate the joint.

Because of the size of the parts involved, and the high activity and contaminating tendency of the flame, plumbing fluxes are typically much more chemically active, and more acidic, than electronic fluxes. Because plumbing joints may be done at any angle, even upside down, plumbing fluxes are generally formulated as pastes which stay in place better than liquids. Flux should be applied to all surfaces of the joint, inside and out. Flux residues should be removed after the joint is complete or they can, eventually, erode through the copper substrates and cause failure of the joint.

Many plumbing solder formulations are available, with different characteristics, such as higher or lower melting temperature, depending on the specific requirements of the job. Building codes currently almost universally require the use of lead-free solder for potable water piping, though traditional tin-lead solder is still available. Studies have shown that lead-soldered plumbing pipes can result in elevated levels of lead in drinking water.

Since copper pipe quickly conducts heat away from a joint, great care must be taken to ensure that the joint is properly heated through to obtain a good bond. After the joint is properly cleaned, fluxed and fitted, the torch flame is applied to the thickest part of the joint, typically the fitting with the pipe inside it, with the solder applied at the gap between the tube and the fitting. When all the parts are heated through, the solder will melt and flow into the joint by capillary action. The torch may need to be moved around the joint to ensure all areas are wetted out. However, the installer must take care to not overheat the areas being soldered. If the tube begins to discolor it means that the tube has been over-heated and is beginning to oxidize, stopping the flow of the solder and causing the soldered joint not to seal properly. Before oxidation the molten solder will follow the heat of the torch around the joint. When the joint is properly wetted out, the solder and then the heat are removed, and while the joint is still very hot, it is usually wiped with a dry rag. This removes excess solder as well as flux residue before it cools down and hardens. With a solder ring joint, the joint is heated until a ring of molten solder is visible around the edge of the fitting and allowed to cool.

Of the three methods of connecting copper tubing, solder connections require the most skill, but soldering copper is a very reliable process, provided some basic conditions are provided:

- The tubing and fittings must be cleaned to bare metal with no tarnish

- Any pressure which is formed by heating of the tubing must have an outlet

- The joint must be dry (which can be challenging when repairing water pipes)

Copper is only one material that is joined in this manner. Brass fittings are often used for valves or as a connection fitting between copper and other metals. Brass piping is soldered in this manner in the making of brass instruments and some woodwind (saxophone and flute) musical instruments

Mechanical and Aluminium Soldering

A number of solder materials, primarily zinc alloys, are used for soldering aluminium metal and alloys and to some lesser extent steel and zinc. This mechanical soldering is similar to a low temperature brazing operation, in that the mechanical characteristics of the joint are reasonably good and it can be used for structural repairs of those materials.

The American welding society defines brazing as using filler metals with melting points over 450 °C (842 °F) — or, by the traditional definition in the United States, above 800 °F (427 °C). Aluminium soldering alloys generally have melting temperatures around 730 °F (388 °C). This soldering / brazing operation can use a propane torch heat source.

These materials are often advertised as "aluminium welding", but the process does not involve melting the base metal, and therefore is not properly a weld.

United States Military Standard or MIL-SPEC specification MIL-R-4208 defines one standard for these zinc-based brazing/soldering alloys. A number of products meet this specification. or very similar performance standards.

Resistance soldering is soldering in which the heat required to flow the solder is created by passing an electric current through the solder. When current is conducted through a resistive material a certain level of heat is generated. By regulating the amount of current conducted and the level of resistance encountered, the amount of heat produced can be predetermined and controlled.

Electrical resistance (usually described as a material's opposition to the flow of an electric current) is used to convert electric energy into thermal energy as an electric current (I) conducted through a material with resistance (R) releases power (P) equal to: $P = I^2 R$, where P is the power measured in watts, I is the current measured in amps and R is the resistance measured in ohms.

Resistance soldering

Resistance soldering is unlike using a conduction iron, where heat is produced within an element and then passed through a thermally conductive tip into the joint area. A cold soldering iron requires time to reach working temperature and must be kept hot between solder joints. Thermal transfer may be inhibited if the tip is not kept properly wetted during use. With resistance soldering an intense heat can be rapidly developed directly within the joint area and in a tightly controlled manner. This allows a faster ramp up to the required solder melt temperature and minimizes thermal travel away from the solder joint, which helps to minimize the potential for thermal damage to materials or components in the surrounding area. Heat is only produced while each joint is being made, making resistance soldering more energy efficient. Resistance soldering equipment, unlike conduction irons, can be used for difficult soldering and brazing applications where significantly higher temperatures may be required. This makes resistance comparable to flame soldering in some situations. When the required temperature can be achieved by either flame or resistance methods the resistance heat is more localized because of direct contact, whereas the flame will spread thus heating a potentially larger area.

Stained Glass Soldering

Historically, stained glass soldering tips were copper, heated by being placed in a charcoal-burning brazier. Multiple tips were used; when one tip cooled down from use, it was placed back in the brazier of charcoal and the next tip was used.

More recently, electrically heated soldering irons are used. These are heated by a coil or ceramic heating element inside the tip of the iron. Different power ratings are available, and temperature can be controlled electronically. These characteristics allow longer beads to be run without interrupting the work to change tips. Soldering irons designed for electronic use are often effective though they are sometimes underpowered for the heavy copper and lead came used in stained glass work. Oleic acid is the classic flux material that has been used to improve solderability.

Tiffany-type stained glass is made by gluing copper foil around the edges of the pieces of glass and then soldering them together. This method makes it possible to create three-dimensional stained glass pieces.

Solderability

The solderability of a substrate is a measure of the ease with which a soldered joint can be made to that material.

Desoldering and Resoldering

Used solder contains some of the dissolved base metals and is unsuitable for reuse in making new joints. Once the solder's capacity for the base metal has been achieved it will no longer properly bond with the base metal, usually resulting in a brittle cold solder joint with a crystalline appearance.

It is good practice to remove solder from a joint prior to resoldering—desoldering braids or vacuum desoldering equipment (solder suckers) can be used. Desoldering wicks contain plenty of flux that will lift the contamination from the copper trace and any device leads that are present. This will leave a bright, shiny, clean junction to be resoldered.

The lower melting point of solder means it can be melted away from the base metal, leaving it mostly intact, though the outer layer will be "tinned" with solder. Flux will remain which can easily be removed by abrasive or chemical processes. This tinned layer will allow solder to flow into a new joint, resulting in a new joint, as well as making the new solder flow very quickly and easily.

Lead-free Electronic Soldering

More recently environmental legislation has specifically targeted the wide use of lead in the electronics industry. The RoHS directives in Europe required many new electronic circuit boards to be lead free by 1 July 2006, mostly in the consumer goods industry, but in some others as well. In Japan lead was phased out prior to legislation by manufacturers due to the additional expense in recycling products containing lead. However, even without the presence of lead, soldering can release fumes that are harmful and/or toxic to humans. It is highly recommended to use a device that can remove the fumes from the work area either by ventilating outside or filtering the air.

It is a common misconception that lead free soldering requires higher soldering temperatures than lead/tin solder; the wetting temperature in lead/tin solder is higher than the melting point and is the controlling factor - Wave soldering can proceed at the same temperature as previous lead/tin soldering. Nevertheless, many new technical challenges have arisen with this endeavor; to reduce the melting point of tin-based solder alloys various new alloys have had to be researched, with additives of copper, silver, bismuth as typical minor additives to reduce melting point and control other properties, additionally tin is a more corrosive metal, and can eventually lead to the failure of solder baths etc.

Lead-free construction has also extended to components, pins, and connectors. Most of these pins used copper frames, and either lead, tin, gold or other finishes. Tin finishes are the most popular of lead-free finishes. Nevertheless, this brings up the issue of how to deal with tin whiskers. The current movement brings the electronics industry back to the problems solved in the 1960s by adding lead. JEDEC has created a classification system to help lead-free electronic manufacturers decide what provisions to take against whiskers, depending upon their application.

Soldering Defects

In the joining of copper tube, failure to properly heat and fill a joint may lead to a 'void' being formed. This is usually a result of improper placement of the flame. If the heat of the flame is not directed at the back of the fitting cup, and the solder wire applied degrees opposite the flame, then solder will quickly fill the opening of the fitting, trapping some flux inside the joint. This bubble of trapped flux is the void; an area inside a soldered joint where solder is unable to completely fill the fittings' cup, because flux has become sealed inside the joint, preventing solder from occupying that space.

Electronics

Various problems may arise in the soldering process which lead to joints which are nonfunctional either immediately or after a period of use.

The most common defect when hand-soldering results from the parts being joined not exceeding the solder's liquidus temperature, resulting in a "cold solder" joint. This is usually the result of the soldering iron being used to heat the solder directly, rather than the parts themselves. Properly done, the iron heats the parts to be connected, which in turn melt the solder, guaranteeing adequate heat in the joined parts for thorough wetting. In electronic hand soldering the flux is embedded in the solder. Therefore, heating the solder first may cause the flux to evaporate before it cleans the surfaces being soldered. A cold-soldered joint may not conduct at all, or may conduct only intermittently. Cold-soldered joints also happen in mass production, and are a common cause of equipment which passes testing, but malfunctions after sometimes years of operation. A "dry joint" occurs when the cooling solder is moved, and often occurs because the joint moves when the soldering iron is removed from the joint.

An improperly selected or applied flux can cause joint failure. If not properly cleaned, a flux may corrode the joint and cause eventual joint failure. Without flux the joint may not be clean, or may be oxidized, resulting in an unsound joint.

In electronics non-corrosive fluxes are often used. Therefore, cleaning flux off may merely be a matter of aesthetics or to make visual inspection of joints easier in specialised 'mission critical' applications such as medical devices, military and aerospace. For satellites, this will also reduce weight, slightly but usefully. In high humidity, even non-corrosive flux might remain slightly active, therefore the flux may be removed to reduce corrosion over time. In some applications, the PCB might also be coated in some form of protective material such as a lacquer to protect it and exposed solder joints from the environment.

Movement of metals being soldered before the solder has cooled will cause a highly unreliable cracked joint. In electronics soldering terminology this is known as a 'dry' joint. It has a characteristically dull or grainy appearance immediately after the joint is made, rather than being smooth, bright and shiny. This appearance is caused by crystallization of the liquid solder. A dry joint is weak mechanically and a poor conductor electrically.

In general a good-looking soldered joint *is* a good joint. As mentioned, it should be smooth, bright, and shiny. If the joint has lumps or balls of otherwise shiny solder the metal has not 'wetted' properly. Not being bright and shiny suggests a weak 'dry' joint. However, technicians trying to apply this guideline when using lead-free solder formulations may experience frustration, because these types of solders readily cool to a dull surface even if the joint is good. The solder looks shiny while molten, and suddenly hazes over as it solidifies even though it has not been disturbed during cooling.

In electronics a 'concave' fillet is ideal. This indicates good wetting and minimal use of solder (therefore minimal *heating* of heat sensitive components). A joint may be good, but if a large amount of unnecessary solder is used then more heating is obviously required. Excessive heating of a PCB may result in 'delamination' - the copper track may actually lift off the board, particularly on single sided PCBs without through hole plating.

Tools

In principle any type of soldering tool can carry out any work using solder at temperatures it can generate. In practice different tools are more suitable for different applications.

Hand-soldering tools widely used for electronics work include the electric soldering iron, which can be fitted with a variety of tips ranging from blunt to very fine, to chisel heads for hot-cutting plastics rather than soldering. The simplest irons do not have temperature regulation; small irons rapidly cool when used to solder to, say, a metal chassis, while large irons have tips too cumbersome for working on PCBs and similar fine work. Temperature-controlled irons have a reserve of power and can maintain temperature over a wide range of work. The soldering gun heats faster but has a larger and heavier body. Gas-powered irons using a catalytic tip to heat a bit, without flame, are used for portable applications. Hot-air guns and pencils allow rework of component packages which cannot easily be performed with electric irons and guns.

For non-electronic applications soldering torches use a flame rather than a soldering tip to heat solder. Soldering torches are often powered by butane and are available in sizes ranging from very small butane/oxygen units suitable for very fine but high-temperature jewelry work, to full-size oxy-fuel torches suitable for much larger work such as copper piping. Common multipurpose propane torches, the same kind used for heat-stripping paint and thawing pipes, can be used for soldering pipes and other fairly large objects either with or without a soldering tip attachment; pipes are generally soldered with a torch by directly applying the open flame.

A soldering copper is a tool with a large copper head and a long handle which is heated in a blacksmith's forge fire and used to apply heat to sheet metal for soldering. Typical soldering coppers have heads weighing between one and four pounds. The head provides a large thermal mass to store enough heat for soldering large areas before needing re-heating in the fire; the larger the head, the longer the working time. Historically, soldering coppers were standard tools used in auto bodywork, although body solder has been mostly superseded by spot welding for mechanical connection, and non-metallic fillers for contouring.

Toaster ovens and hand held infrared lights have been used by hobbyists to replicate production soldering processes on a much smaller scale.

Bristle brushes are usually used to apply plumbing paste flux. For electronic work, flux-core solder is generally used, but additional flux may be used from a flux pen or dispensed from a small bottle with a syringe-like needle.

Wire brush, wire wool and emery cloth are commonly used to prepare plumbing joints for connection. Electronic joints are usually made between surfaces that have been tinned and rarely require mechanical cleaning, though tarnished component leads and copper traces with a dark layer of oxide passivation (due to aging), as on a new prototyping board that has been on the shelf for about a year or more, may need to be mechanically cleaned.

Some fluxes for electronics are designed to be stable and inactive when cool and do not need to be cleaned off, though they still can be if desired, while other fluxes are acidic and must be removed after soldering to prevent corrosion of the circuits. For PCB assembly and rework, either an alcohol or acetone is commonly used with cotton swabs or bristle brushes to remove flux residue after soldering. A heavy rag is usually used to remove flux from a plumbing joint before it cools and hardens. A fiberglass brush can also be used.

A heat sink, such as a crocodile clip, can be used to prevent damaging heat-sensitive components while hand-soldering. The heat sink limits the temperature of the component body by absorbing and dissipating heat (reducing the thermal resistance between the component and the air), while the thermal resistance of the leads maintains the temperature difference between the part of the leads being soldered and the component body so that the leads become hot enough to melt the solder while the component body remains cooler.

Dip Soldering

Dip soldering is a small-scale soldering process by which electronic components are soldered to a printed circuit board (PCB) to form an electronic assembly. The solder wets to the exposed metallic areas of the board (those not protected with solder mask), creating a reliable mechanical and electrical connection.

Dip soldering apparatus.

Dip soldering is used for both through-hole printed circuit assemblies, and surface mount. It is one of the cheapest methods to solder and is extensively used in the small scale industries of developing countries .

Dip soldering is the manual equivalent of automated wave soldering. The apparatus required is just a small tank containing molten solder. PCB with mounted components is dipped manually into the tank when the molten solder sticks to the exposed metallic areas of the board.

Dip Solder Process

Dip soldering is accomplished by submerging parts to be joined into a molten solder bath. Thus, all components surfaces are coated with filler metal. Solders have low surface tension and high wetting capability. There are many types of solders, each used for different applications. Such as Lead-Silver for strength at higher than room temperature. Tin-Lead is used for General Purpose; Tin-Zinc is used for Aluminum; Cadmium-Silver is used for strength at high temperatures; Zinc-Aluminum is used for Aluminum and corrosion resistance; Tin-Silver and Tin-Bismuth is used for Electronics. Because of the toxicity of lead, lead-free solders are being developed and more widely used. The molten bath can be any suitable filler metal, but the selection is usually confined to the lower melting point elements. The most common dip soldering operations use zinc-aluminum and tin-lead solders. Solder pot metal - Cast iron or steel,electrically heated. Bath temperature - 220 deg. Celsius to 260 deg.Celsius (for binary tin-lead alloys) Bath temperature - 350 deg.Celsius to 400 deg. Celsius (for lead-rich alloys)Solder composition - 60% Sn(tin),40% Pb(lead) or eutectic alloy.

Process Schematic

The workpieces to be joined are treated with cleaning flux. Then the workpiece is mounted in the workholding device and immersed in the molten solder for 2 to 12 seconds. The workpiece is often agitated to aid the flow of the solder. The workpiece holder must allow an inclination of 3 to 5 deg. so that the solder may run off to insure a smooth finish.

Workpiece Geometry

This process is generally limited to all metal work pieces, although other materials, such as circuit boards can also tolerate momentary contact with the hot molten solder without damage.

Setup and Equipment

There is not much equipment or setup for this process, all that is needed is the solder pot with its temperature control panel, the bath of molten solder, and the work holding device. Usually the work holding device is custom made for each respective workpiece for either manual or automated dipping.

Solderability

Some materials are easier to solder than others. Copper, silver, and gold are easy to solder. Iron and Nickel are a little more difficult. Titanium, magnesium, cast irons, steels, ceramics, and graphites are hard to solder. However, if they are first plated they are more easily soldered. An example of this is tin-plating, in which a steel is sheet coated with tin so that it can be soldered more easily.

Applications

Dip Soldering is used extensively in the electronics industry. However, they have a limited service use at elevated temperatures because of the low melting point of the filler metals. Soldered materials do not have much strength and are therefore not used for load-bearing.

Ultrasonic Soldering

Ultrasonic soldering is a flux-less soldering process that uses ultrasonic energy, without the need for chemicals to solder materials, such as glass, ceramics, and composites, hard to solder metals and other sensitive components which cannot be soldered using conventional means. Ultrasonic (U/S) soldering, as a flux-less soldering process, is finding growing application in soldering of metals and ceramics from solar photovoltaics and medical shape memory alloys to specialized electronic and sensor packages. U/S soldering has been reported since 1955 as a method to solder aluminum and other metals without the use of flux.

Ultrasonic soldering is a distinctly different process than ultrasonic welding. Ultrasonic welding uses ultrasonic energy to join parts without adding any kind of filler material while ordinary soldering uses external heating to melt filler metal materials, namely solders, to form a joint. Ultrasonic soldering can be done with either a specialized soldering iron or a specialized solder pot. In either case the process can be automated for large-scale production or can be done by hand for prototyping or repair work. Initially, U/S soldering was aimed at joining aluminum and other metals; however, with the emergence of active solders, a much wider range of metals, ceramics and glass can now be soldered.

Ultrasonic soldering uses either ultrasonically coupled heated solder iron tips (0.5 – 10 mm) or ultrasonically coupled solder baths as mentioned above. In these devices, piezoelectric crystals are used to generate high frequency (20 – 60 kHz) acoustic waves in molten solder layers or batch, to mechanically disrupt oxides that form on the molten solder surfaces. The tips for U/S soldering irons are also coupled to a heating element while the piezoelectric crystal is thermally isolated, not to degrade the piezoelectric element. Ultrasonic soldering iron tips can heat (up to 450 °C) while mechanically oscillating at 20 – 60 kHz. This soldering tip can melt solder filler metals as acoustic vibrations are induced in the molten solder pool. The vibration and cavitation in the molten solder then permits solders to wet and adhere to many metal surfaces.

The acoustic energy created by the solder tip or ultrasonic solder pot works via cavitation of the molten solder which mechanically disrupts oxide layers on the solder layers themselves and on metal surfaces being joined.

Cavitation in the molten solder pool can be very effective in disrupting the oxides on many metals, however, it is not effective when soldering to ceramics and glass since they themselves are oxides or other non-metal compound that cannot be disrupted since they are the base materials. In the cases of soldering direct to glasses and ceramics, ultrasonic soldering filler metals need to be modified with active elements such as In, Ti, Hf, Zr and rare earth elements (Ce, La, and Lu). Solders when alloyed with these elements are called "active solders" since they directly act on the glass/ceramic surfaces to create a bond.

The use of ultrasonic soldering is expanding, since it is clean and flux-less in combination with active solders being specified for joining assemblies where either corrosive flux can be trapped or otherwise disrupt operation or contaminate clean production environments or there are dissimilar materials / metals / ceramic/ glasses being joined. To be effective in adhering to surfaces, active solders' own nascent oxide on melting need to be disrupted and ultrasonic agitation is well suited.

In applications where the area of the solder joint is a small or band, U/S soldering using 1 – 10 mm tips can be very effective since the volume of molten metal is small and can effectively be agitated by the 1 – 10 mm U/S soldering iron tips. The figures in this article show the U/S soldering equipment (power supply and soldering tools-tips) and the application of solder to glass using U/S solder iron tips. In other larger surface bonding application, as shown in the image below, wide, heated U/S tips are being used to spread and wet active solders on large aluminum surfaces (and is applicable to other metal, ceramic and glass surfaces.

Wave Soldering

Selective soldering machine

Wave soldering is a bulk soldering process used in the manufacture of printed circuit boards. The circuit board is passed over a pan of molten solder in which a pump produces an upwelling of solder that looks like a standing wave. As the circuit board makes contact with this wave, the components become soldered to the board. Wave soldering is used for both through-hole printed circuit assemblies, and surface mount. In the latter case, the components are glued onto the surface of a printed circuit board (PCB) by placement equipment, before being run through the molten solder wave.

As through-hole components have been largely replaced by surface mount components, wave soldering has been supplanted by reflow soldering methods in many large-scale electronics applications. However, there is still significant wave soldering where surface-mount technology (SMT) is not suitable (e.g., large power devices and high pin count connectors), or where simple through-hole technology prevails (certain major appliances).

Wave Solder Process

A simple wave soldering machine.

There are many types of wave solder machines; however, the basic components and principles of these machines are the same. The basic equipment used during the process is a conveyor that moves the PCB through the different zones, a pan of solder used in the soldering process, a pump that produces the actual wave, the sprayer for the flux and the preheating pad. The solder is usually a mixture of metals. A typical solder has the chemical makeup of 50% tin, 49.5% lead, and 0.5% antimony.

Fluxing

Flux in the wave soldering process has a primary and a secondary objective. The primary objective is to clean the components that are to be soldered, principally any oxide layers that may have formed. There are two types of flux, corrosive and noncorrosive. Noncorrosive flux requires precleaning and is used when low acidity is required. Corrosive flux is quick and requires little precleaning, but has a higher acidity.

Preheating

Preheating helps to accelerate the soldering process and to prevent thermal shock.

Cleaning

Some types of flux, called "no-clean" fluxes, do not require cleaning; their residues are benign after the soldering process. Typically no-clean fluxes are especially sensitive to process conditions, which may make them undesirable in some applications. Other kinds of flux, however, require a cleaning stage, in which the PCB is washed with solvents and/or deionized water to remove flux residue.

Finish and Quality

Quality depends on proper temperatures when heating and on properly treated surfaces.

Defect	Possible causes	Effects
Cracks	Mechanical Stress	Loss of Conductivity
Cavities	Contaminated surface	Reduction in strength
	Lack of flux	Poor conductivity
	Insufficient preheating	
Wrong solder thickness	Wrong solder temperature	Susceptible to stress
	Wrong conveyor speed	Too thin for current load
		Undesired bridging between paths
Poor Conductor	Contaminated solder	Product Failures

Solder Types

Different combinations of tin, lead and other metals are used to create solder. The combinations used depend on the desired properties. The most popular combinations are SAC (Tin(Sn)/Silver(Ag)/Copper(Cu)) alloys and Sn63Pb37 (Sn63A) which is 63% tin, 37% lead. The latter combination is strong, has a low melting range, and melts and sets quickly. Higher tin compositions give the solder higher corrosion resistances, but raise the melting point. Another common composition is 11% tin, 37% lead, 42% bismuth, and 10% cadmium. This combination has a low melting point and is useful for soldering components that are sensitive to heat. Environmental and performance requirements also factor into alloy selection. Common restrictions include restrictions on lead (Pb) when RoHS compliance is required and restrictions on pure tin (Sn) when long term reliability is a concern.

Effects of Cooling Rate

It is important that the PCBs be allowed to cool at a reasonable rate. If they are cooled too fast, then the PCB can become warped and the solder can be compromised. On the other hand, if the PCB is allowed to cool too slowly, then the PCB can become brittle and some components may be damaged by heat. The PCB should be cooled by either a fine water spray or air cooled to decrease the amount of damage to the board.

Thermal Profiling

Thermal profiling is the act of measuring several points on a circuit board to determine the thermal excursion it takes through the soldering process. In the electronics manufacturing industry, SPC (Statistical Process Control) helps determine if the process is in control, measured against the reflow parameters defined by the soldering technologiEs and component requirements. Products like the Solderstar WaveShuttle and the Optiminer have been developed special fixtures which are passed through the process and can measure the temperature profile, along with contact times, wave parallelism and wave heights. These fixture combined with analysis software allows the production engineer to establish and then control the wave solder process.

Butt Welding

Butt welding is a welding technique used to connect parts which are nearly parallel and don't overlap. It can be used to run a processing machine continuously, as opposed to having to restart such machine with a new supply of metals. Butt-welding is an economical and reliable way of joining without using additional components.

Usually, a butt-welding joint is made by gradually heating up the two weld ends with a weld plate and then joining them under a specific pressure. This process is very suitable for prefabrication and producing special fittings. Afterward, the material is usually ground down to a smooth finish and either sent on its way to the processing machine, or sold as a completed product.

This type of weld is usually accomplished with an arc or MIG welder. It can also be accomplished by brazing. With arc welding, after the butt weld is complete, the weld itself needs to be struck with a hammer forge to remove slag (a type of waste material) before any subsequent welds can be applied. This is not necessary for MIG welds however, as a protective gas removes any need for slag to appear. Another with a MIG welder is that a continuous copper coated wire is fed onto the stock, making the weld virtually inexhaustible.

Hand Welding

A joint between two members aligned approximately in the same plane. Butt welding can also be achieved through traditional blow torches in the most common form of butt joints, a process that uses some variety of flux, usually a tin-based solder and precise hand-eye coordination that is common for hand-made boxes of copper, brass, and silver. There are two types of butt welding; one is carried out by smiting and another is carried out by welding two work pieces by non-overlapping.

The process consists of two desired strips of metal that are lined with flux that is lightly dried with a blowtorch until it is a sticky consistency, followed by cutting a strip of solder that is generally 20% of the full joint's size. Applying heat gently makes the gel-like flux now appear white and powdery which now is primed to be welded in which the blow torch is arched so that the "heat cone", the bluest and hottest part of the flame, is now directly upon the solder melting the joints together evenly.

The joint is then cooled and cleaned in a solution of sulfuric acid diluted in 20 parts water – commonly known as "pickle" – to remove imperfections. Sanding and polishing then achieves the desired finishing.

Upset Weld

The parts to be welded are clamped edge to edge in copper jaws of the welding machine and brought together in a solid contact so that their point of contact forms a locality of high electric resistance, while current flows to heat the joint . At this point the pressure applied upsets or forges the parts together . Upset buttwelding is used principally on non ferrous materials for welding bars, rods, wire, tubing, formed parts, etc.

Standards

EN 1993-1-8, which covers the design of joints in the design of steel structures, defines a set of provisions for welding structural steel.

References

- AWS A3.0:2001, Standard Welding Terms and Definitions Including Terms for Adhesive Bonding, Brazing, Soldering, Thermal Cutting, and Thermal Spraying, American Welding Society (2001), p. 118. ISBN 0-87171-624-0

- Shea, William R., ed. (1983). Nature mathematized: historical and philosophical case studies in classical modern natural philosophy. Dordrecht: Reidel. p. 282. ISBN 978-90-277-1402-2.

- Houldcroft, P. T. (1973) [1967]. "Chapter 3: Flux-Shielded Arc Welding". Welding Processes. Cambridge University Press. p. 23. ISBN 0-521-05341-2.

- Carlisle, Rodney (2004). Scientific American Inventions and Discoveries, p.365. John Wright & Songs, Inc., New Jersey. ISBN 0-471-24410-4.

- William Augustus Tilden. Chemical Discovery and Invention in the Twentieth Century. Adamant Media Corporation. p. 80. ISBN 0-543-91646-4.

- Cary, Howard B; Scott C. Helzer (2005). Modern Welding Technology. Upper Saddle River, New Jersey: Pearson Education. ISBN 0-13-113029-3.

- Joseph R. Davis, ASM International. Handbook Committee (2001). Copper and copper alloys. ASM International. p. 311. ISBN 0-87170-726-8.

- AWS A3.0:2001, Standard Welding Terms and Definitions Including Terms for Adhesive Bonding, Brazing, Soldering, Thermal Cutting, and Thermal Spraying, American Welding Society (2001), p. 118. ISBN 0-87171-624-0

- "Testing of work environments for electromagnetic interference". Pacing Clin Electrophysiol. 15 (11 Pt 2): 2016–22. 1992. doi:10.1111/j.1540-8159.1992.tb03013.x. PMID 1279591.

- Supplies of Cadmium Bearing Silver Solders Continue (2009-01-20). "Strength of Silver Solder Joints". www.cupalloys.co.uk. Retrieved 2010-07-26.

Metals used in Welding

Welding is the art of joining two or more metals, this requires using numerous metals during the process. Some of these metals are aluminum, titanium, beryllium and zirconium. This section closely examines all the metals used in welding and provides an easy understanding of the subject matter.

Aluminium

Aluminium or aluminum (in North American English) is a chemical element in the boron group with symbol Al and atomic number 13. It is a silvery-white, soft, nonmagnetic, ductile metal. Aluminium is the third most abundant element in the Earth's crust (after oxygen and silicon) and its most abundant metal. Aluminium makes up about 8% of the crust by mass, though it is less common in the mantle below. Aluminium metal is so chemically reactive that native specimens are rare and limited to extreme reducing environments. Instead, it is found combined in over 270 different minerals. The chief ore of aluminium is bauxite.

Aluminium is remarkable for the metal's low density and its ability to resist corrosion through the phenomenon of passivation. Aluminium and its alloys are vital to the aerospace industry and important in transportation and structures, such as building facades and window frames. The oxides and sulfates are the most useful compounds of aluminium.

Despite its prevalence in the environment, no known form of life uses aluminium salts metabolically, but aluminium is well tolerated by plants and animals. Because of their abundance, the potential for a biological role is of continuing interest and studies continue.

Characteristics

"Bauxite tailings" storage facility in Stade, Germany. The aluminium industry generates about 70 million tons of this waste annually.

Physical

Aluminium is a relatively soft, durable, lightweight, ductile, and malleable metal with appearance ranging from silvery to dull gray, depending on the surface roughness. It is nonmagnetic and does not easily ignite. A fresh film of aluminium serves as a good reflector (approximately 92%) of visible light and an excellent reflector (as much as 98%) of medium and far infrared radiation. The yield strength of pure aluminium is 7–11 MPa, while aluminium alloys have yield strengths ranging from 200 MPa to 600 MPa. Aluminium has about one-third the density and stiffness of steel. It is easily machined, cast, drawn and extruded.

Aluminium atoms are arranged in a face-centered cubic (fcc) structure. Aluminium has a stacking-fault energy of approximately 200 mJ/m^2.

Aluminium is a good thermal and electrical conductor, having 59% the conductivity of copper, both thermal and electrical, while having only 30% of copper's density. Aluminium is capable of superconductivity, with a superconducting critical temperature of 1.2 kelvin and a critical magnetic field of about 100 gauss (10 milliteslas).

Chemical

Corrosion resistance can be excellent because a thin surface layer of aluminium oxide forms when the bare metal is exposed to air, effectively preventing further oxidation, in a process termed passivation. The strongest aluminium alloys are less corrosion resistant due to galvanic reactions with alloyed copper. This corrosion resistance is greatly reduced by aqueous salts, particularly in the presence of dissimilar metals.

In highly acidic solutions, aluminium reacts with water to form hydrogen, and in highly alkaline ones to form aluminates— protective passivation under these conditions is negligible. Primarily because it is corroded by dissolved chlorides, such as common sodium chloride, household plumbing is never made from aluminium.

However, because of its general resistance to corrosion, aluminium is one of the few metals that retains silvery reflectance in finely powdered form, making it an important component of silver-colored paints. Aluminium mirror finish has the highest reflectance of any metal in the 200–400 nm (UV) and the 3,000–10,000 nm (far IR) regions; in the 400–700 nm visible range it is slightly outperformed by tin and silver and in the 700–3000 nm (near IR) by silver, gold, and copper.

Aluminium is oxidized by water at temperatures below 280 °C to produce hydrogen, aluminium hydroxide and heat:

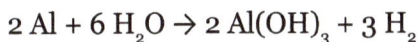

$$2\,Al + 6\,H_2O \rightarrow 2\,Al(OH)_3 + 3\,H_2$$

This conversion is of interest for the production of hydrogen. However, commercial application of this fact has challenges in circumventing the passivating oxide layer, which inhibits the reaction, and in storing the energy required to regenerate the aluminium metal.

Isotopes

Aluminium has many known isotopes, with mass numbers range from 21 to 42; however, only

^{27}Al (stable) and ^{26}Al (radioactive, $t_{1/2}$ = 7.2×10^5 years) occur naturally. ^{27}Al has a natural abundance above 99.9%. ^{26}Al is produced from argon in the atmosphere by spallation caused by cosmic-ray protons. Aluminium isotopes are useful in dating marine sediments, manganese nodules, glacial ice, quartz in rock exposures, and meteorites. The ratio of ^{26}Al to ^{10}Be has been used to study transport, deposition, sediment storage, burial times, and erosion on 10^5 to 10^6 year time scales. Cosmogenic ^{26}Al was first applied in studies of the Moon and meteorites. Meteoroid fragments, after departure from their parent bodies, are exposed to intense cosmic-ray bombardment during their travel through space, causing substantial ^{26}Al production. After falling to Earth, atmospheric shielding drastically reduces ^{26}Al production, and its decay can then be used to determine the meteorite's terrestrial age. Meteorite research has also shown that ^{26}Al was relatively abundant at the time of formation of our planetary system. Most meteorite scientists believe that the energy released by the decay of ^{26}Al was responsible for the melting and differentiation of some asteroids after their formation 4.55 billion years ago.

Natural Occurrence

Stable aluminium is created when hydrogen fuses with magnesium, either in large stars or in supernovae. It is estimated to be the 14th most common element in the Universe, by mass-fraction. However, among the elements that have odd atomic numbers, aluminium is the third most abundant by mass fraction, after hydrogen and nitrogen.

In the Earth's crust, aluminium is the most abundant (8.3% by mass) metallic element and the third most abundant of all elements (after oxygen and silicon). The Earth's crust has a greater abundance of aluminium than the rest of the planet, primarily in aluminium silicates. In the Earths mantle, which is only 2% aluminium by mass, these aluminium silicate minerals are largely replaced by silica and magnesium oxides. Overall, the Earth is about 1.4% aluminium by mass (eighth in abundance by mass). Aluminium occurs in greater proportion in the Earth than in the Solar system and Universe because the more common elements (hydrogen, helium, neon, nitrogen, carbon as hydrocarbon) are volatile at Earth's proximity to the Sun and large quantities of those were lost.

Because of its strong affinity for oxygen, aluminium is almost never found in the elemental state; instead it is found in oxides or silicates. Feldspars, the most common group of minerals in the Earth's crust, are aluminosilicates. Native aluminium metal can only be found as a minor phase in low oxygen fugacity environments, such as the interiors of certain volcanoes. Native aluminium has been reported in cold seeps in the northeastern continental slope of the South China Sea. Chen *et al.* (2011) propose the theory that these deposits resulted from bacterial reduction of tetrahydroxoaluminate $Al(OH)_4^-$.

Aluminium also occurs in the minerals beryl, cryolite, garnet, spinel, and turquoise. Impurities in Al_2O_3, such as chromium and iron, yield the gemstones ruby and sapphire, respectively.

Although aluminium is a common and widespread element, not all aluminium minerals are economically viable sources of the metal. Almost all metallic aluminium is produced from the ore bauxite ($AlO_x(OH)_{3-2x}$). Bauxite occurs as a weathering product of low iron and silica bedrock in tropical climatic conditions. Bauxite is mined from large deposits in Australia, Brazil, Guinea, and Jamaica; it is also mined from lesser deposits in China, India, Indonesia, Russia, and Suriname.

Production and Refinement

Bauxite, a major aluminium ore. The red-brown color is due to the presence of iron minerals.

Bayer Process and Hall–Héroult Processes

Bauxite is converted to aluminium oxide (Al_2O_3) by the Bayer process. Relevant chemical equations are:

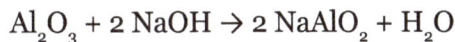

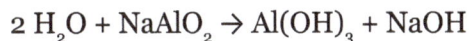

$$Al_2O_3 + 2\,NaOH \rightarrow 2\,NaAlO_2 + H_2O$$

$$2\,H_2O + NaAlO_2 \rightarrow Al(OH)_3 + NaOH$$

The intermediate, sodium aluminate, with the simplified formula $NaAlO_2$, is soluble in strongly alkaline water, and the other components of the ore are not. Depending on the quality of the bauxite ore, twice as much waste ("Bauxite tailings") as alumina is generated.

The conversion of alumina to aluminium metal is achieved by the Hall–Héroult process. In this energy-intensive process, a solution of alumina in a molten (950 and 980 °C (1,740 and 1,800 °F)) mixture of cryolite (Na_3AlF_6) with calcium fluoride is electrolyzed to produce metallic aluminium:

$$Al^{3+} + 3\,e^- \rightarrow Al$$

The liquid aluminium metal sinks to the bottom of the solution and is tapped off, and usually cast into large blocks called aluminium billets for further processing. Carbon dioxide is produced at the carbon anode:

$$2\,O^{2-} + C \rightarrow CO_2 + 4\,e^-$$

The carbon anode is consumed by reaction with oxygen to form carbon dioxide gas, with a small quantity of fluoride gases. In modern smelters, the gas is filtered through alumina to remove fluorine compounds and return aluminium fluoride to the electrolytic cells. The anode the reduction cell must be replaced regularly, since it is consumed in the process. The cathode is also eroded, mainly by electrochemical processes and liquid metal movement induced by intense electrolytic currents. After five to ten years, depending on the current used in the electrolysis, a cell must be rebuilt because of cathode wear.

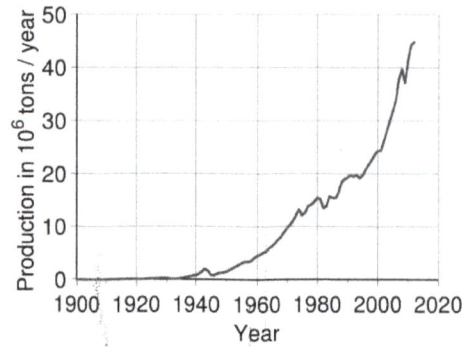

World production trend of aluminium

Aluminium electrolysis with the Hall–Héroult process consumes a lot of energy. The worldwide average specific energy consumption is approximately 15±0.5 kilowatt-hours per kilogram of aluminium produced (52 to 56 MJ/kg). Some smelters achieve approximately 12.8 kW·h/kg (46.1 MJ/kg). (Compare this to the heat of reaction, 31 MJ/kg, and the Gibbs free energy of reaction, 29 MJ/kg.) Minimizing line currents for older technologies are typically 100 to 200 kiloamperes; state-of-the-art smelters operate at about 350 kA.

The Hall–Heroult process produces aluminium with a purity of above 99%. Further purification can be done by the Hoopes process. This process involves the electrolysis of molten aluminium with a sodium, barium and aluminium fluoride electrolyte. The resulting aluminium has a purity of 99.99%.

Electric power represents about 20% to 40% of the cost of producing aluminium, depending on the location of the smelter. Aluminium production consumes roughly 5% of electricity generated in the U.S. Aluminium producers tend to locate smelters in places where electric power is both plentiful and inexpensive—such as the United Arab Emirates with its large natural gas supplies, and Iceland and Norway with energy generated from renewable sources. The world's largest smelters of alumina are located in the People's Republic of China, Russia and the provinces of Quebec and British Columbia in Canada.

Aluminium spot price 1987–2012

In 2005, the People's Republic of China was the top producer of aluminium with almost a one-fifth world share, followed by Russia, Canada, and the US, reports the British Geological Survey.

Over the last 50 years, Australia has become the world's top producer of bauxite ore and a major producer and exporter of alumina (before being overtaken by China in 2007). Australia produced

77 million tonnes of bauxite in 2013. The Australian deposits have some refining problems, some being high in silica, but have the advantage of being shallow and relatively easy to mine.

Aluminium Chloride Electrolysis Process

The high energy consumption of Hall–Héroult process motivated the development of the electrolytic process based on aluminium chloride. The pilot plant with 6500 tons/year output was started in 1976 by Alcoa. The plant offered two advantages: (i) energy requirements were 40% less than plants using the Hall–Héroult process, and (ii) the more accessible kaolinite (instead of bauxite and cryolite) was used for feedstock. Nonetheless, the pilot plant was shut down. The reasons for failure were the cost of aluminium chloride, general technology maturity problems, and leakage of the trace amounts of toxic polychlorinated biphenyl compounds.

Aluminium chloride process can also be used for the co-production of titanium, depending on titanium contents in kaolinite.

Aluminium Carbothermic Process

The non-electrolytic *aluminium carbothermic* process of aluminium production would theoretically be cheaper and consume less energy. However, it has been in the experimental phase for decades because the high operating temperature creates difficulties in material technology that have not yet been solved.

Recycling

Aluminium recycling code

Aluminium is theoretically 100% recyclable without any loss of its natural qualities. According to the International Resource Panel's Metal Stocks in Society report, the global per capita stock of aluminium in use in society (i.e. in cars, buildings, electronics etc.) is 80 kg (180 lb). Much of this is in more-developed countries (350–500 kg (770–1,100 lb) per capita) rather than less-developed countries (35 kg (77 lb) per capita). Knowing the per capita stocks and their approximate lifespans is important for planning recycling.

Recovery of the metal through recycling has become an important task of the aluminium industry. Recycling was a low-profile activity until the late 1960s, when the growing use of aluminium beverage cans brought it to public awareness.

Recycling involves melting the scrap, a process that requires only 5% of the energy used to produce aluminium from ore, though a significant part (up to 15% of the input material) is lost as dross (ash-like oxide). An aluminium stack melter produces significantly less dross, with values reported below 1%. The dross can undergo a further process to extract aluminium.

Europe has achieved high rates of aluminium recycling ranging from 42% of beverage cans, 85% of construction materials, and 95% of transport vehicles.

Recycled aluminium is known as secondary aluminium, but maintains the same physical properties as primary aluminium. Secondary aluminium is produced in a wide range of formats and is employed in 80% of alloy injections. Another important use is extrusion.

White dross from primary aluminium production and from secondary recycling operations still contains useful quantities of aluminium that can be extracted industrially. The process produces aluminium billets, together with a highly complex waste material. This waste is difficult to manage. It reacts with water, releasing a mixture of gases (including, among others, hydrogen, acetylene, and ammonia), which spontaneously ignites on contact with air; contact with damp air results in the release of copious quantities of ammonia gas. Despite these difficulties, the waste is used as a filler in asphalt and concrete.

Compounds

Oxidation state +3

The vast majority of compounds, including all Al-containing minerals and all commercially significant aluminium compounds, feature aluminium in the oxidation state 3+. The coordination number of such compounds varies, but generally Al^{3+} is six-coordinate or tetracoordinate. Almost all compounds of aluminium(III) are colorless.

Halides

All four trihalides are well known. Unlike the structures of the three heavier trihalides, aluminium fluoride (AlF_3) features six-coordinate Al. The octahedral coordination environment for AlF_3 is related to the compactness of the fluoride ion, six of which can fit around the small Al^{3+} center. AlF_3 sublimes (with cracking) at 1,291 °C (2,356 °F). With heavier halides, the coordination numbers are lower. The other trihalides are dimeric or polymeric with tetrahedral Al centers. These materials are prepared by treating aluminium metal with the halogen, although other methods exist. Acidification of the oxides or hydroxides affords hydrates. In aqueous solution, the halides often form mixtures, generally containing six-coordinate Al centers that feature both halide and aquo ligands. When aluminium and fluoride are together in aqueous solution, they readily form complex ions such as [AlF(H2O)5]2+, AlF3(H2O)3, and [AlF6]3−. In the case of chloride, polyaluminium clusters are formed such as $[Al_{13}O_4(OH)_{24}(H_2O)_{12}]^{7+}$.

Oxide and Hydroxides

Aluminium forms one stable oxide, known by its mineral name corundum. Sapphire and ruby are impure corundum contaminated with trace amounts of other metals. The two oxide-hydroxides, AlO(OH), are boehmite and diaspore. There are three trihydroxides: bayerite, gibbsite, and nordstrandite, which differ in their crystalline structure (polymorphs). Most are produced from ores by a variety of wet processes using acid and base. Heating the hydroxides leads to formation of corundum. These materials are of central importance to the production of aluminium and are themselves extremely useful.

Carbide, Nitride, and Related Materials

Aluminium carbide (Al_4C_3) is made by heating a mixture of the elements above 1,000 °C (1,832 °F). The pale yellow crystals consist of tetrahedral aluminium centers. It reacts with water or dilute acids to give methane. The acetylide, $Al_2(C_2)_3$, is made by passing acetylene over heated aluminium.

Aluminium nitride (AlN) is the only nitride known for aluminium. Unlike the oxides, it features tetrahedral Al centers. It can be made from the elements at 800 °C (1,472 °F). It is air-stable material with a usefully high thermal conductivity. Aluminium phosphide (AlP) is made similarly; it hydrolyses to give phosphine:

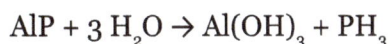

$$AlP + 3\ H_2O \rightarrow Al(OH)_3 + PH_3$$

Organoaluminium Compounds and Related Hydrides

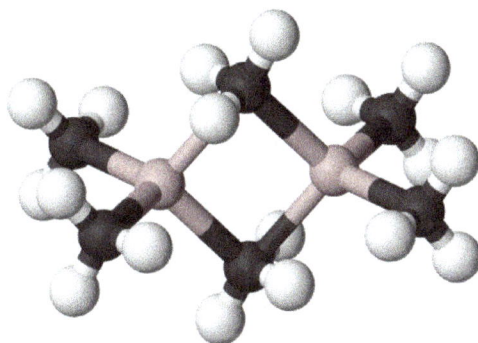

Structure of trimethylaluminium, a compound that features five-coordinate carbon.

A variety of compounds of empirical formula AlR_3 and $AlR_{1.5}Cl_{1.5}$ exist. These species usually feature tetrahedral Al centers formed by dimerization with some R or Cl bridging between both Al atoms, e.g. "trimethylaluminium" has the formula $Al_2(CH_3)_6$ (see figure). With large organic groups, triorganoaluminium compounds exist as three-coordinate monomers, such as triisobutylaluminium. Such compounds are widely used in industrial chemistry, despite the fact that they are often highly pyrophoric. Few analogues exist between organoaluminium and organoboron compounds other than large organic groups.

The important aluminium hydride is lithium aluminium hydride ($LiAlH_4$), which is used in as a reducing agent in organic chemistry. It can be produced from lithium hydride and aluminium trichloride:

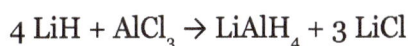

$$4\ LiH + AlCl_3 \rightarrow LiAlH_4 + 3\ LiCl$$

Several useful derivatives of $LiAlH_4$ are known, e.g. sodium bis(2-methoxyethoxy)dihydridoaluminate. The simplest hydride, aluminium hydride or alane, remains a laboratory curiosity. It is a polymer with the formula $(AlH_3)_n$, in contrast to the corresponding boron hydride that is a dimer with the formula $(BH_3)_2$.

Oxidation states +1 and +2

Although the great majority of aluminium compounds feature Al^{3+} centers, compounds with lower oxidation states are known and sometime of significance as precursors to the Al^{3+} species.

Aluminium(I)

AlF, AlCl and AlBr exist in the gaseous phase when the trihalide is heated with aluminium. The composition AlI is unstable at room temperature, converting to triiodide:

$$3\,AlI -> AlI3 + 2\,Al$$

A stable derivative of aluminium monoiodide is the cyclic adduct formed with triethylamine, Al$_4$I$_4$(NEt$_3$)$_4$. Also of theoretical interest but only of fleeting existence are Al$_2$O and Al$_2$S. Al$_2$O is made by heating the normal oxide, Al$_2$O$_3$, with silicon at 1,800 °C (3,272 °F) in a vacuum. Such materials quickly disproportionate to the starting materials.

Aluminium(II)

Very simple Al(II) compounds are invoked or observed in the reactions of Al metal with oxidants. For example, aluminium monoxide, AlO, has been detected in the gas phase after explosion and in stellar absorption spectra. More thoroughly investigated are compounds of the formula R$_4$Al$_2$ which contain an Al-Al bond and where R is a large organic ligand.

Analysis

The presence of aluminium can be detected in qualitative analysis using aluminon.

Applications

Etched surface from a high purity (99.9998%) aluminium bar, size 55×37 mm

General Use

Aluminium is the most widely used non-ferrous metal. Global production of aluminium in 2005 was 31.9 million tonnes. It exceeded that of any other metal except iron (837.5 million tonnes). Forecast for 2012 was 42–45 million tonnes, driven by rising Chinese output.

Aluminium is almost always alloyed, which markedly improves its mechanical properties, especially when tempered. For example, the common aluminium foils and beverage cans are alloys of 92% to 99% aluminium. The main alloying agents are copper, zinc, magnesium, manganese, and silicon (e.g., duralumin) with the levels of other metals in a few percent by weight.

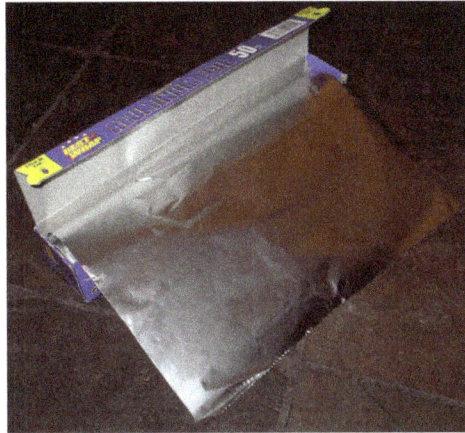

Household aluminium foil

Aluminium-bodied Austin *"A40 Sports"* (c. 1951)

Aluminium slabs being transported from a smelter

Some of the many uses for aluminium metal are in:

- Transportation (automobiles, aircraft, trucks, railway cars, marine vessels, bicycles, spacecraft, etc.) as sheet, tube, and castings.

- Packaging (cans, foil, frame of etc.).

- Food and beverage containers, because of its resistance to corrosion.

- Construction (windows, doors, siding, building wire, sheathing, roofing, etc.).

- A wide range of household items, from cooking utensils to baseball bats and watches.

- Street lighting poles, sailing ship masts, walking poles.

- Outer shells and cases for consumer electronics and photographic equipment.

- Electrical transmission lines for power distribution ("creep" and oxidation are not issues in this application as the terminations are usually multi-sided "crimps" which enclose all sides of the conductor with a gas-tight seal).

- MKM steel and Alnico magnets.

- Super purity aluminium (SPA, 99.980% to 99.999% Al), used in electronics and CDs, and also in wires/cabling.

- Heat sinks for transistors, CPUs, and other components in electronic appliances.

- Substrate material of metal-core copper clad laminates used in high brightness LED lighting.

- Light reflective surfaces and paint.

- Pyrotechnics, solid rocket fuels, and thermite.

- Production of hydrogen gas by reaction with hydrochloric acid or sodium hydroxide.

- In alloy with magnesium to make aircraft bodies and other transportation components.

- Cooking utensils, because of its resistant to corrosion and light-weight.

- Coins in such countries as France, Italy, Poland, Finland, Romania, Israel, and the former Yugoslavia struck from aluminium or an aluminium-copper alloy.

- Musical instruments. Some guitar models sport aluminium diamond plates on the surface of the instruments, usually either chrome or black. Kramer Guitars and Travis Bean are both known for having produced guitars with necks made of aluminium, which gives the instrument a very distinctive sound. Aluminium is used to make some guitar resonators and some electric guitar speakers.

Aluminium is usually alloyed – it is used as pure metal only when corrosion resistance and/or workability is more important than strength or hardness. The strength of aluminium alloys is abruptly increased with small additions of scandium, zirconium, or hafnium. A thin layer of aluminium can be deposited onto a flat surface by physical vapor deposition or (very infrequently) chemical vapor deposition or other chemical means to form optical coatings and mirrors.

Aluminium Compounds

Because aluminium is abundant and most of its derivatives exhibit low toxicity, the compounds of aluminium enjoy wide and sometimes large-scale applications.

Alumina

Aluminium oxide (Al_2O_3) and the associated oxy-hydroxides and trihydroxides are produced or extracted from minerals on a large scale. The great majority of this material is converted to metallic aluminium. In 2013, about 10% of the domestic shipments in the United States were used for other applications. One major use is to absorb water where it is viewed as a contaminant or impurity. Alumina is used to remove water from hydrocarbons in preparation for subsequent processes that would be poisoned by moisture.

Aluminium oxides are common catalysts for industrial processes; e.g. the Claus process to convert hydrogen sulfide to sulfur in refineries and to alkylate amines. Many industrial catalysts are "supported" by alumina, meaning that the expensive catalyst material (e.g., platinum) is dispersed over a surface of the inert alumina.

Being a very hard material (Mohs hardness 9), alumina is widely used as an abrasive; being extraordinarily chemically inert, it is useful in highly reactive environments such as high pressure sodium lamps.

Sulfates

Several sulfates of aluminium have industrial and commercial application. Aluminium sulfate ($Al_2(SO_4)_3 \cdot (H_2O)_{18}$) is produced on the annual scale of several billions of kilograms. About half of the production is consumed in water treatment. The next major application is in the manufacture of paper. It is also used as a mordant, in fire extinguishers, in fireproofing, as a food additive (E number E173), and in leather tanning. Aluminium ammonium sulfate, which is also called ammonium alum, $(NH_4)Al(SO_4)_2 \cdot 12H_2O$, is used as a mordant and in leather tanning, as is aluminium potassium sulfate ($[Al(K)](SO_4)_2) \cdot (H_2O)_{12}$. The consumption of both alums is declining.

Chlorides

Aluminium chloride ($AlCl_3$) is used in petroleum refining and in the production of synthetic rubber and polymers. Although it has a similar name, aluminium chlorohydrate has fewer and very different applications, particularly as a colloidal agent in water purification and an antiperspirant. It is an intermediate in the production of aluminium metal.

Niche Compounds

Many aluminium compounds have niche applications:

- Aluminium acetate in solution is used as an astringent.

- Aluminium borate ($Al_2O_3 \cdot B_2O_3$) and aluminium fluorosilicate ($Al_2(SiF_6)_3$) are used in the production of glass, ceramics, synthetic gemstones.

- Aluminium phosphate ($AlPO_4$) used in the manufacture of glass, ceramic, pulp and paper products, cosmetics, paints, varnishes, and in dental cement.

- Aluminium hydroxide ($Al(OH)_3$) is used as an antacid, and mordant; it is used also in water purification, the manufacture of glass and ceramics, and in the waterproofing fabrics.

- Lithium aluminium hydride is a powerful reducing agent used in organic chemistry.

- Organoaluminiums are used as Lewis acids and cocatalysts.

- Methylaluminoxane is a cocatalyst for Ziegler-Natta olefin polymerization to produce vinyl polymers such as polyethene.

- Aqueous aluminium ions (such as aqueous aluminium sulfate) are used to treat against fish parasites such as *Gyrodactylus salaris*.

- In many vaccines, certain aluminium salts serve as an immune adjuvant (immune response booster) to allow the protein in the vaccine to achieve sufficient potency as an immune stimulant.

Aluminium Alloys in Structural Applications

Aluminium foam

Aluminium alloys with a wide range of properties are used in engineering structures. Alloy systems are classified by a number system (ANSI) or by names indicating their main alloying constituents (DIN and ISO).

The strength and durability of aluminium alloys vary widely, not only as a result of the components of the specific alloy, but also as a result of heat treatments and manufacturing processes. A lack of knowledge of these aspects has from time to time led to improperly designed structures and gained aluminium a bad reputation.

One important structural limitation of aluminium alloys is their fatigue strength. Unlike steels, aluminium alloys have no well-defined fatigue limit, meaning that fatigue failure eventually occurs, under even very small cyclic loadings. Engineers must assess applications and design for a fixed and finite life of the structure, rather than infinite life.

Another important property of aluminium alloys is sensitivity to heat. Workshop procedures are complicated by the fact that aluminium, unlike steel, melts without first glowing red. Manual blow torch operations require additional skill and experience. Aluminium alloys, like all structural alloys, are subject to internal stresses after heat operations such as welding and casting. The lower melting points of aluminium alloys make them more susceptible to distortions from thermally induced stress relief. Stress can be relieved and controlled during manufacturing by heat-treating the parts in an oven, followed by gradual cooling—in effect annealing the stresses.

The low melting point of aluminium alloys has not precluded use in rocketry, even in combustion chambers where gases can reach 3500 K. The Agena upper stage engine used regeneratively cooled aluminium in some parts of the nozzle, including the thermally critical throat region.

Another alloy of some value is aluminium bronze (Cu-Al alloy).

History

The statue of Anteros in Piccadilly Circus, London, was made in 1893 and is one of the first statues cast in aluminium.

Although ancient Greeks and Romans used aluminium salts as dyeing mordants and as astringents for dressing wounds, metallic aluminum was not refined until the modern era. Alum, a salt of aluminum and potassium, is still used as a styptic. In 1782, Guyton de Morveau suggested calling the "base" of (i.e., the metallic element in) alum *alumine*. In 1808, Humphry Davy identified the existence of a metal base of alum, which he at first termed *alumium* and later *aluminum*.

The metal was first produced in 1825 in an impure form by Danish physicist and chemist Hans Christian Ørsted. He reacted anhydrous aluminium chloride with potassium amalgam, yielding a lump of metal looking similar to tin. Friedrich Wöhler was aware of these experiments and cited them, but after repeating Ørsted's experiments, he concluded that this metal was pure potassium. He conducted a similar experiment in 1827 by mixing anhydrous aluminium chloride with potassium and produced aluminium. Wöhler is therefore generally credited with isolating aluminium. Further, Pierre Berthier discovered aluminium in bauxite ore. Henri Etienne Sainte-Claire Deville improved Wöhler's method in 1846. As described in his 1859 book, aluminium trichloride could be reduced by sodium, which was more convenient and less expensive than potassium, which Wöhler had used. In the mid-1880s, aluminium metal was exceedingly difficult to produce, which made pure aluminium more valuable than gold. So celebrated was the metal that bars of aluminium were exhibited at the Exposition Universelle of 1855. Napoleon III of France is reputed to have held a banquet where the most honored guests were given aluminium utensils, while the others made do with gold.

Aluminium was selected as the material to use for the 100 ounces (2.8 kg) capstone of the Washington Monument in 1884, a time when one ounce (30 grams) cost the daily wage of a common worker on the project (in 1884 about $1 for 10 hours of labor; today, a construction worker in the US working on such a project might earn $25–$35 per hour and therefore around $300 in an equivalent single 10-hour day). The capstone, which was set in place on 6 December 1884 in an elaborate dedication ceremony, was the largest single piece of aluminium cast at the time.

The Cowles companies supplied aluminium alloy in quantity in the United States and England using smelters like the furnace of Carl Wilhelm Siemens by 1886.

Hall-Heroult Process: Availability of Cheap Aluminium Metal

Charles Martin Hall of Ohio in the U.S. and Paul Héroult of France independently developed the Hall-Héroult electrolytic process that facilitated large-scale production of metallic aluminium. This process remains in use today. In 1888, with the financial backing of Alfred E. Hunt, the Pittsburgh Reduction Company started; today it is known as Alcoa. Héroult's process was in production by 1889 in Switzerland at Aluminium Industrie, now Alcan, and at British Aluminium, now Luxfer Group and Alcoa, by 1896 in Scotland.

By 1895, the metal was being used as a building material as far away as Sydney, Australia in the dome of the Chief Secretary's Building.

With the explosive expansion of the airplane industry during World War I (1914–1917), major governments demanded large shipments of aluminium for light, strong airframes. They often subsidized factories and the necessary electrical supply systems.

Many navies have used an aluminium superstructure for their vessels; the 1975 fire aboard USS *Belknap* that gutted her aluminium superstructure, as well as observation of battle damage to British ships during the Falklands War, led to many navies switching to all steel superstructures.

Aluminium wire was once widely used for domestic electrical wiring in the United States, and a number of fires resulted from creep and corrosion-induced failures at junctions and terminations; additional and preventable factors in the failures have been identified. Aluminium is still used in electrical services with specially designed wire termination hardware.

Etymology

The International Union of Pure and Applied Chemistry (IUPAC) adopted *aluminium* as the standard international name for the element in 1990 but, three years later, recognized *aluminum* as an acceptable variant. The IUPAC periodic table uses the *aluminium* spelling only. IUPAC internal publications use the two spelling with nearly equal frequency.

Different Endings

Most countries use the ending "-ium" for "aluminium". In the United States and Canada, the ending "-um" predominates. The Canadian Oxford Dictionary prefers *aluminum*, whereas the Australian Macquarie Dictionary prefers *aluminium*. In 1926, the American Chemical Society officially decided to use *aluminum* in its publications; American dictionaries typically label the spelling *aluminium* as "chiefly British". The earliest citation given in the Oxford English Dictionary for any word used as a name for this element is *alumium*, which British chemist and inventor Humphry Davy employed in 1808 for the metal he was trying to isolate electrolytically from the mineral *alumina*. The citation is from the journal *Philosophical Transactions of the Royal Society of London*: "Had I been so fortunate as to have obtained more certain evidences on this subject, and to have procured the metallic substances I was in search of, I should have proposed for them the names of silicium, alumium, zirconium, and glucium."

Davy settled on *aluminum* by the time he published his 1812 book *Chemical Philosophy*: "This substance appears to contain a peculiar metal, but as yet Aluminum has not been obtained in a perfectly free state, though alloys of it with other metalline substances have been procured sufficiently distinct to indicate the probable nature of alumina." But the same year, an anonymous contributor to the *Quarterly Review,* a British political-literary journal, in a review of Davy's book, objected to *aluminum* and proposed the name *aluminium*, "for so we shall take the liberty of writing the word, in preference to aluminum, which has a less classical sound."

The -*ium* suffix followed the precedent set in other newly discovered elements of the time: potassium, sodium, magnesium, calcium, and strontium (all of which Davy isolated himself). Nevertheless, element names ending in -*um* were not unknown at the time; for example, platinum (known to Europeans since the 16th century), molybdenum (discovered in 1778), and tantalum (discovered in 1802). The -*um* suffix is consistent with the universal spelling alumina for the oxide (as opposed to aluminia), as lanthana is the oxide of lanthanum, and magnesia, ceria, and thoria are the oxides of magnesium, cerium, and thorium respectively.

The *aluminum* spelling is used in the Webster's Dictionary of 1828. In his advertising handbill for his new electrolytic method of producing the metal in 1892, Charles Martin Hall used the -*um* spelling, despite his constant use of the -*ium* spelling in all the patents he filed between 1886 and 1903. Hall's domination of production of the metal ensured that *aluminum* became the standard English spelling in North America.

Biology

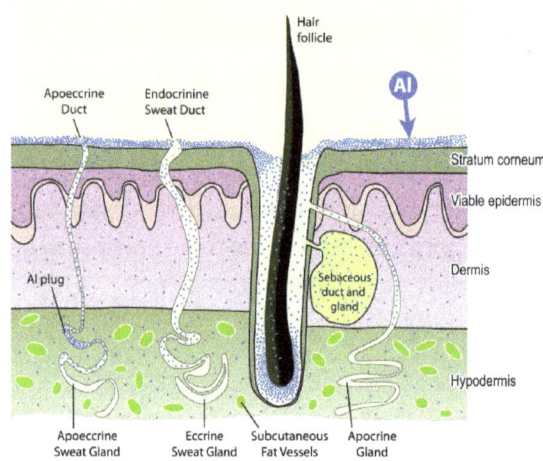

Schematic of Al absorption by human skin.

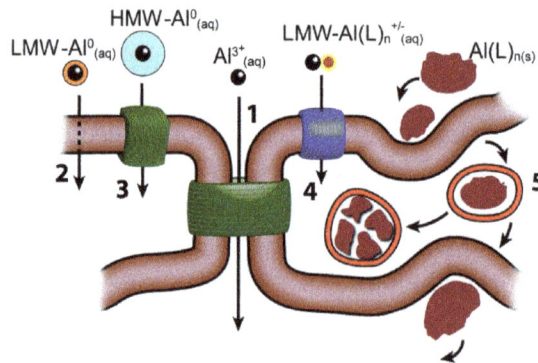

There are five major Al forms absorbed by human body: the free solvated trivalent cation ($Al^{3+}_{(aq)}$); low-molecular-weight, neutral, soluble complexes ($LMW\text{-}Al^0_{(aq)}$); high-molecular-weight, neutral, soluble complexes ($HMW\text{-}Al^0_{(aq)}$); low-molecular-weight, charged, soluble complexes ($LMW\text{-}Al(L)_n^{+/-}{}_{(aq)}$); nano and micro-particulates ($Al(L)_{n(s)}$). They are transported across cell membranes or cell epi-/endothelia through five major routes: (1) paracellular; (2) transcellular; (3) active transport; (4) channels; (5) adsorptive or receptor-mediated endocytosis.

Despite its widespread occurrence in the Earth crust, aluminium has no known function in biology. Aluminium salts are remarkably nontoxic, aluminium sulfate having an LD_{50} of 6207 mg/kg (oral, mouse), which corresponds to 500 grams for an 80 kg (180 lb) person. The extremely low acute toxicity notwithstanding, the health effects of aluminium are of interest in view of the widespread occurrence of the element in the environment and in commerce.

Health Concerns

In very high doses, aluminium is associated with altered function of the blood–brain barrier. A small percentage of people are allergic to aluminium and experience contact dermatitis, digestive disorders, vomiting or other symptoms upon contact or ingestion of products containing aluminium, such as antiperspirants and antacids. In those without allergies, aluminium is not as toxic as heavy metals, but there is evidence of some toxicity if it is consumed in amounts greater than

40 mg/day per kg of body mass. The use of aluminium cookware has not been shown to lead to aluminium toxicity in general, however excessive consumption of antacids containing aluminium compounds and excessive use of aluminium-containing antiperspirants provide more significant exposure levels. Consumption of acidic foods or liquids with aluminium enhances aluminium absorption, and maltol has been shown to increase the accumulation of aluminium in nerve and bone tissues. Aluminium increases estrogen-related gene expression in human breast cancer cells cultured in the laboratory. The estrogen-like effects of these salts have led to their classification as metalloestrogens.

There is little evidence that aluminium in antiperspirants causes skin irritation. Nonetheless, its occurrence in antiperspirants, dyes (such as aluminium lake), and food additives has caused concern. Although there is little evidence that normal exposure to aluminium presents a risk to healthy adults, some studies point to risks associated with increased exposure to the metal. Aluminium in food may be absorbed more than aluminium from water. It is classified as a non-carcinogen by the US Department of Health and Human Services.

In case of suspected sudden intake of a large amount of aluminium, deferoxamine mesylate may be given to help eliminate it from the body by chelation.

Occupational Safety

Exposure to powdered aluminium or aluminium welding fumes can cause pulmonary fibrosis. The United States Occupational Safety and Health Administration (OSHA) has set a permissible exposure limit of 15 mg/m³ time weighted average (TWA) for total exposure and 5 mg/m³ TWA for respiratory exposure. The US National Institute for Occupational Safety and Health (NIOSH) recommended exposure limit is the same for respiratory exposure but is 10 mg/m³ for total exposure, and 5 mg/m³ for fumes and powder.

Fine aluminium powder can ignite or explode, posing another workplace hazard.

Alzheimer's Disease

Aluminium has controversially been implicated as a factor in Alzheimer's disease. According to the Alzheimer's Society, the medical and scientific opinion is that studies have not convincingly demonstrated a causal relationship between aluminium and Alzheimer's disease. Nevertheless, some studies, such as those on the PAQUID cohort, cite aluminium exposure as a risk factor for Alzheimer's disease. Some brain plaques have been found to contain increased levels of the metal. Research in this area has been inconclusive; aluminium accumulation may be a consequence of the disease rather than a causal agent.

Effect on Plants

Aluminium is primary among the factors that reduce plant growth on acid soils. Although it is generally harmless to plant growth in pH-neutral soils, the concentration in acid soils of toxic Al^{3+} cations increases and disturbs root growth and function.

Most acid soils are saturated with aluminium rather than hydrogen ions. The acidity of the soil is therefore, a result of hydrolysis of aluminium compounds. The concept of "corrected lime potential" is now used to define the degree of base saturation in soil testing to determine the "lime requirement".

Wheat has developed a tolerance to aluminium, releasing of organic compounds that bind to harmful aluminium cations. Sorghum is believed to have the same tolerance mechanism. The first gene for aluminium tolerance has been identified in wheat. It was shown that sorghum's aluminium tolerance is controlled by a single gene, as for wheat. This adaptation is not found in all plants.

Biodegradation

A Spanish scientific report from 2001 claimed that the fungus *Geotrichum candidum* consumes the aluminium in compact discs. Other reports all refer back to the 2001 Spanish report and there is no supporting original research. Better documented, the bacterium *Pseudomonas aeruginosa* and the fungus *Cladosporium resinae* are commonly detected in aircraft fuel tanks that use kerosene-based fuels (not AV gas), and laboratory cultures can degrade aluminium. However, these life forms do not directly attack or consume the aluminium; rather, the metal is corroded by microbe waste products.

Titanium

Titanium is a chemical element with symbol Ti and atomic number 22. It is a lustrous transition metal with a silver color, low density and high strength. It is highly resistant to corrosion in sea water, aqua regia, and chlorine.

Titanium was discovered in Cornwall, Great Britain, by William Gregor in 1791 and named by Martin Heinrich Klaproth for the Titans of Greek mythology. The element occurs within a number of mineral deposits, principally rutile and ilmenite, which are widely distributed in the Earth's crust and lithosphere, and it is found in almost all living things, rocks, water bodies, and soils. The metal is extracted from its principal mineral ores by the Kroll and Hunter processes. The most common compound, titanium dioxide, is a popular photocatalyst and is used in the manufacture of white pigments. Other compounds include titanium tetrachloride ($TiCl_4$), a component of smoke screens and catalysts; and titanium trichloride ($TiCl_3$), which is used as a catalyst in the production of polypropylene.

Titanium can be alloyed with iron, aluminium, vanadium, and molybdenum, among other elements, to produce strong, lightweight alloys for aerospace (jet engines, missiles, and spacecraft), military, industrial process (chemicals and petro-chemicals, desalination plants, pulp, and paper), automotive, agri-food, medical prostheses, orthopedic implants, dental and endodontic instruments and files, dental implants, sporting goods, jewelry, mobile phones, and other applications.

The two most useful properties of the metal are corrosion resistance and the highest strength-to-density ratio of any metallic element. In its unalloyed condition, titanium is as strong as some steels, but less dense. There are two allotropic forms and five naturally occurring isotopes of this element, ^{46}Ti through ^{50}Ti, with ^{48}Ti being the most abundant (73.8%). Although they have the same number of valence electrons and are in the same group in the periodic table, titanium and zirconium differ in many chemical and physical properties.

Characteristics

Physical Properties

A metallic element, titanium is recognized for its high strength-to-weight ratio. It is a strong metal with low density that is quite ductile (especially in an oxygen-free environment), lustrous, and metallic-white in color. The relatively high melting point (more than 1,650 °C or 3,000 °F) makes it useful as a refractory metal. It is paramagnetic and has fairly low electrical and thermal conductivity.

Commercial (99.2% pure) grades of titanium have ultimate tensile strength of about 434 MPa (63,000 psi), equal to that of common, low-grade steel alloys, but are less dense. Titanium is 60% denser than aluminium, but more than twice as strong as the most commonly used 6061-T6 aluminium alloy. Certain titanium alloys (e.g., Beta C) achieve tensile strengths of over 1400 MPa (200000 psi). However, titanium loses strength when heated above 430 °C (806 °F).

Titanium is not as hard as some grades of heat-treated steel, is non-magnetic and a poor conductor of heat and electricity. Machining requires precautions, because the material might gall if sharp tools and proper cooling methods are not used. Like those made from steel, titanium structures have a fatigue limit that guarantees longevity in some applications. Titanium alloys have less stiffness than many other structural materials such as aluminium alloys and carbon fiber.

The metal is a dimorphic allotrope of an hexagonal α form that changes into a body-centered cubic (lattice) β form at 882 °C (1,620 °F). The specific heat of the α form increases dramatically as it is heated to this transition temperature but then falls and remains fairly constant for the β form regardless of temperature. Similar to zirconium and hafnium, an additional omega phase exists, which is thermodynamically stable at high pressures, but is metastable at ambient pressures. This phase is usually hexagonal (*ideal*) or trigonal (*distorted*) and can be considered to be due to a soft longitudinal acoustic phonon of the β phase causing collapse of (111) planes of atoms.

Chemical Properties

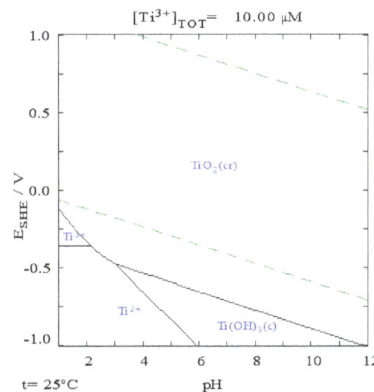

The Pourbaix diagram for titanium in pure water, perchloric acid or sodium hydroxide

Like aluminium and magnesium, titanium metal and its alloys oxidize immediately upon exposure to air. Titanium readily reacts with oxygen at 1,200 °C (2,190 °F) in air, and at 610 °C (1,130 °F) in pure oxygen, forming titanium dioxide. It is, however, slow to react with water and air at ambient temperatures because it forms a passive oxide coating that protects the bulk metal from further oxidation. When it first forms, this protective layer is only 1–2 nm thick but continues to grow slowly; reaching a thickness of 25 nm in four years.

Atmospheric passivation gives titanium excellent resistance to corrosion, almost equivalent to platinum, capable of withstanding attack by dilute sulfuric and hydrochloric acids, chloride solutions, and most organic acids. However, titanium is corroded by concentrated acids. As indicated by its negative redox potential, titanium is thermodynamically a very reactive metal that burns in normal atmosphere at lower temperatures than the melting point. Melting is possible only in an inert atmosphere or in a vacuum. At 550 °C (1,022 °F), it combines with chlorine. It also reacts with the other halogens and absorbs hydrogen.

Titanium is one of the few elements that burns in pure nitrogen gas, reacting at 800 °C (1,470 °F) to form titanium nitride, which causes embrittlement. Because of its high reactivity with oxygen, nitrogen, and some other gases, titanium filaments are applied in titanium sublimation pumps as scavengers for these gases. Such pumps inexpensively and reliably produce extremely low pressures in ultra-high vacuum systems.

Occurrence

2011 production of rutile and ilmenite		
Country	thousand tonnes	% of total
Australia	1300	19.4
South Africa	1160	17.3
Canada	700	10.4
India	574	8.6
Mozambique	516	7.7
China	500	7.5
Vietnam	490	7.3
Ukraine	357	5.3
World	**6700**	**100**

Titanium is the ninth-most abundant element in Earth's crust (0.63% by mass) and the seventh-most abundant metal. It is present as oxides in most igneous rocks, in sediments derived from them, in living things, and natural bodies of water. Of the 801 types of igneous rocks analyzed by the United States Geological Survey, 784 contained titanium. Its proportion in soils is approximately 0.5 to 1.5%.

It is widely distributed and occurs primarily in the minerals anatase, brookite, ilmenite, perovskite, rutile and titanite (sphene). Of these minerals, only rutile and ilmenite have economic importance, yet even they are difficult to find in high concentrations. About 6.0 and 0.7 million tonnes of those minerals were mined in 2011, respectively. Significant titanium-bearing ilmenite deposits exist in western Australia, Canada, China, India, Mozambique, New Zealand, Norway, Ukraine and South Africa. About 186,000 tonnes of titanium metal sponge were produced in 2011, mostly in China (60,000 t), Japan (56,000 t), Russia (40,000 t), United States (32,000 t) and Kazakhstan (20,700 t). Total reserves of titanium are estimated to exceed 600 million tonnes.

The concentration of Ti is about 4 picomolar in the ocean. At 100 °C, the concentration of titanium in water is estimated to be less than 10^{-7} M at pH 7. The identity of titanium species in aqueous solution remains unknown because of its low solubility and the lack of sensitive spectroscopic methods, although only the 4+ oxidation state is stable in air. No evidence exists for a biological role, although rare organisms are known to accumulate high concentrations of titanium.

Titanium is contained in meteorites and has been detected in the Sun and in M-type stars (the coolest type) with a surface temperature of 3,200 °C (5,790 °F). Rocks brought back from the Moon during the Apollo 17 mission are composed of 12.1% TiO_2. It is also found in coal ash, plants, and even the human body. Native titanium (pure metallic) is very rare.

Isotopes

Naturally occurring titanium is composed of 5 stable isotopes: ^{46}Ti, ^{47}Ti, ^{48}Ti, ^{49}Ti, and ^{50}Ti, with ^{48}Ti being the most abundant (73.8% natural abundance). Eleven radioisotopes have been characterized, the most stable being ^{44}Ti with a half-life of 63 years; ^{45}Ti, 184.8 minutes; ^{51}Ti, 5.76 minutes; and ^{52}Ti, 1.7 minutes. All the other radioactive isotopes have half-lives less than 33 seconds and the majority, less than half a second.

The isotopes of titanium range in atomic weight from 39.99 u (^{40}Ti) to 57.966 u (^{58}Ti). The primary decay mode before the most abundant stable isotope, ^{48}Ti, is electron capture and the primary mode after is beta emission. The primary decay products before ^{48}Ti are element 21 (scandium) isotopes and the primary products after are element 23 (vanadium) isotopes.

Titanium becomes radioactive upon bombardment with deuterons, emitting mainly positrons and hard gamma rays.

Compounds

TiN-coated drill bit

The +4 oxidation state dominates titanium chemistry, but compounds in the +3 oxidation state are also common. Commonly, titanium adopts an octahedral coordination geometry in its complexes, but tetrahedral $TiCl_4$ is a notable exception. Because of its high oxidation state, titanium(IV) compounds exhibit a high degree of covalent bonding. Unlike most other transition metals, simple aquo Ti(IV) complexes are unknown.

Oxides, Sulfides, and Alkoxides

The most important oxide is TiO_2, which exists in three important polymorphs; anatase, brookite, and rutile. All of these are white diamagnetic solids, although mineral samples can appear dark. They adopt polymeric structures in which Ti is surrounded by six oxide ligands that link to other Ti centers.

Titanates usually refer to titanium(IV) compounds, as represented by barium titanate ($BaTiO_3$). With a perovskite structure, this material exhibits piezoelectric properties and is used as a transducer in the interconversion of sound and electricity. Many minerals are titanates, e.g. ilmenite ($FeTiO_3$). Star sapphires and rubies get their asterism (star-forming shine) from the presence of titanium dioxide impurities.

A variety of reduced oxides of titanium are known. Ti_3O_5, described as a Ti(IV)-Ti(III) species, is a purple semiconductor produced by reduction of TiO_2 with hydrogen at high temperatures, and is used industrially when surfaces need to be vapour-coated with titanium dioxide: it evaporates as pure TiO, whereas TiO_2 evaporates as a mixture of oxides and deposits coatings with variable refractive index. Also known is Ti_2O_3, with the corundum structure, and TiO, with the rock salt structure, although often nonstoichiometric.

The alkoxides of titanium(IV), prepared by reacting $TiCl_4$ with alcohols, are colourless compounds that convert to the dioxide on reaction with water. They are industrially useful for depositing solid TiO_2 via the sol-gel process. Titanium isopropoxide is used in the synthesis of chiral organic compounds via the Sharpless epoxidation.

Titanium forms a variety of sulfides, but only TiS_2 has attracted significant interest. It adopts a layered structure and was used as a cathode in the development of lithium batteries. Because Ti(IV) is a "hard cation", the sulfides of titanium are unstable and tend to hydrolyze to the oxide with release of hydrogen sulfide.

Nitrides, Carbides

Titanium(III) compounds are characteristically violet, illustrated by this aqueous solution of titanium trichloride.

Titanium nitride (TiN) has a hardness equivalent to sapphire and carborundum (9.0 on the Mohs Scale), and is often used to coat cutting tools, such as drill bits. It is also used as a gold-colored decorative finish and as a barrier metal in semiconductor fabrication. Titanium carbide, which is also very hard, is found in cutting tools and coatings.

Halides

Titanium tetrachloride (titanium(IV) chloride, $TiCl_4$) is a colorless volatile liquid (commercial samples are yellowish) that, in air, hydrolyzes with spectacular emission of white clouds. Via the Kroll process, $TiCl_4$ is produced in the conversion of titanium ores to titanium dioxide, e.g., for use in white paint. It is widely used in organic chemistry as a Lewis acid, for example in the Mukaiyama aldol condensation. In the van Arkel process, titanium tetraiodide (TiI_4) is generated in the production of high purity titanium metal.

Titanium(III) and titanium(II) also form stable chlorides. A notable example is titanium(III) chloride ($TiCl_3$), which is used as a catalyst for production of polyolefins and a reducing agent in organic chemistry.

Organometallic Complexes

Owing to the important role of titanium compounds as polymerization catalyst, compounds with Ti-C bonds have been intensively studied. The most common organotitanium complex is titanocene dichloride ((C_5H_5)$_2TiCl_2$). Related compounds include Tebbe's reagent and Petasis reagent. Titanium forms carbonyl complexes, e.g. (C_5H_5)$_2Ti(CO)_2$.

History

Martin Heinrich Klaproth named titanium for the Titans of Greek mythology.

Titanium was discovered as an inclusion of a mineral in Cornwall, Great Britain, in 1791 by the clergyman and amateur geologist William Gregor, then vicar of Creed parish. He recognized the presence of a new element in ilmenite when he found black sand by a stream in the nearby parish of Manaccan and noticed the sand was attracted by a magnet. Analyzing the sand, he determined the presence of two metal oxides: iron oxide (explaining the attraction to the magnet) and 45.25% of a white metallic oxide he could not identify. Realizing that the unidentified oxide contained a metal that did not match any known element, Gregor reported his findings to the Royal Geological Society of Cornwall and in the German science journal *Crell's Annalen*.

Around the same time, Franz-Joseph Müller von Reichenstein produced a similar substance, but could not identify it. The oxide was independently rediscovered in 1795 by Prussian chemist Martin Heinrich Klaproth in rutile from Boinik (German name of unknown place) village of Hungary (now in Slovakia). Klaproth found that it contained a new element and named it for the Titans of Greek mythology. After hearing about Gregor's earlier discovery, he obtained a sample of manaccanite and confirmed it contained titanium.

The currently known processes for extracting titanium from its various ores are laborious and costly; it is not possible to reduce the ore by heating with carbon (as in iron smelting) because titanium combines with the carbon to produce titanium carbide. Pure metallic titanium (99.9%) was first prepared in 1910 by Matthew A. Hunter at Rensselaer Polytechnic Institute by heating $TiCl_4$ with sodium at 700–800 °C under great pressure in a batch process known as the Hunter process. Titanium metal was not used outside the laboratory until 1932 when William Justin Kroll proved that it could be produced by reducing titanium tetrachloride ($TiCl_4$) with calcium. Eight years later he refined this process with magnesium and even sodium in what became known as the Kroll process. Although research continues into more efficient and cheaper processes (e.g., FFC Cambridge, Armstrong), the Kroll process is still used for commercial production.

Titanium sponge, made by the Kroll process

Titanium of very high purity was made in small quantities when Anton Eduard van Arkel and Jan Hendrik de Boer discovered the iodide, or crystal bar, process in 1925, by reacting with iodine and decomposing the formed vapors over a hot filament to pure metal.

In the 1950s and 1960s the Soviet Union pioneered the use of titanium in military and submarine applications (Alfa class and Mike class) as part of programs related to the Cold War. Starting in the early 1950s, titanium came into use extensively in military aviation, particularly in high-performance jets, starting with aircraft such as the F100 Super Sabre and Lockheed A-12 and SR-71.

Recognizing the strategic importance of titanium the U.S. Department of Defense supported early efforts of commercialization.

Throughout the period of the Cold War, titanium was considered a strategic material by the U.S. government, and a large stockpile of titanium sponge was maintained by the Defense National Stockpile Center, which was finally depleted in the 2000s. According to 2006 data, the world's largest producer, Russian-based VSMPO-Avisma, was estimated to account for about 29% of the world market share. As of 2015, titanium sponge metal was produced in six countries: China, Japan, Russia, Kazakhstan, the USA, Ukraine and India. (in order of output).

In 2006, the U.S. Defense Advanced Research Projects Agency (DARPA) awarded $5.7 million to a two-company consortium to develop a new process for making titanium metal powder. Under heat and pressure, the powder can be used to create strong, lightweight items ranging from armor plating to components for the aerospace, transport, and chemical processing industries.

Production and Fabrication

Titanium (mineral concentrate)

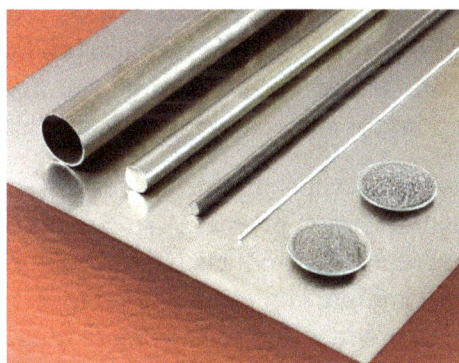

Basic titanium products: plate, tube, rods and powder

The processing of titanium metal occurs in 4 major steps: reduction of titanium ore into "sponge", a porous form; melting of sponge, or sponge plus a master alloy to form an ingot; primary fabrication, where an ingot is converted into general mill products such as billet, bar, plate, sheet, strip, and tube; and secondary fabrication of finished shapes from mill products.

Because it cannot be readily produced by reduction of its dioxide, titanium metal is obtained by reduction of $TiCl_4$ with magnesium metal in the Kroll Process. The complexity of this batch production in the Kroll process explains the relatively high market value of titanium, despite the Kroll process is less expensive than the Hunter process. To produce the $TiCl_4$ required by the Kroll process, the dioxide is subjected to carbothermic reduction in the presence of chlorine. In this process, the chlorine gas is passed over a red-hot mixture of rutile or ilmenite in the presence of carbon. After extensive purification by fractional distillation, the TiCl4 is reduced with 800 °C molten magnesium in an argon atmosphere. Titanium metal can be further purified by the van Arkel–de Boer process, which involves thermal decomposition of titanium tetraiodide.

A more recently developed batch production method, the FFC Cambridge process, consumes titanium dioxide powder (a refined form of rutile) as feedstock and produces titanium metal, either powder or sponge. The process involves fewer steps than the Kroll process and takes less time. If mixed oxide powders are used, the product is an alloy.

Common titanium alloys are made by reduction. For example, cuprotitanium (rutile with copper added is reduced), ferrocarbon titanium (ilmenite reduced with coke in an electric furnace), and manganotitanium (rutile with manganese or manganese oxides) are reduced.

$$2 \, FeTiO_3 + 7 \, Cl_2 + 6 \, C \rightarrow 2 \, TiCl_4 + 2 \, FeCl_3 + 6 \, CO \, (900 \, °C)$$

$$TiCl_4 + 2 \, Mg \rightarrow 2 \, MgCl_2 + Ti \, (1100 \, °C)$$

About 50 grades of titanium and titanium alloys are designed and currently used, although only a couple of dozen are readily available commercially. The ASTM International recognizes 31 Grades of titanium metal and alloys, of which Grades 1 through 4 are commercially pure (unalloyed). Those four vary in tensile strength as a function of oxygen content, with Grade 1 being the most ductile (lowest tensile strength with an oxygen content of 0.18%), and Grade 4 the least ductile (highest tensile strength with an oxygen content of 0.40%). The remaining grades are alloys, each designed for specific properties of ductility, strength, hardness, electrical resistivity, creep resistance, specific corrosion resistance, and combinations thereof.

In addition to the ASTM specifications, titanium alloys are also produced to meet Aerospace and Military specifications (SAE-AMS, MIL-T), ISO standards, and country-specific specifications, as well as proprietary end-user specifications for aerospace, military, medical, and industrial applications.

Titanium powder is manufactured using a flow production process known as the Armstrong process that is similar to the batch production Hunter process. A stream of titanium tetrachloride gas is added to a stream of molten sodium metal; the products (sodium chloride salt and titanium particles) is filtered from the extra sodium. Titanium is then separated from the salt by water washing. Both sodium and chlorine are recycled to produce and process more titanium tetrachloride.

All welding of titanium must be done in an inert atmosphere of argon or helium to shield it from contamination with atmospheric gases (oxygen, nitrogen, and hydrogen). Contamination causes a variety of conditions, such as embrittlement, which reduces the integrity of the assembly welds and leads to joint failure.

Commercially pure flat product (sheet, plate) can be formed readily, but processing must take into account the fact that the metal has a "memory" and tends to spring back. This is especially true of certain high-strength alloys. Titanium cannot be soldered without first pre-plating it in a metal that is solderable. The metal can be machined with the same equipment and the same processes as stainless steel.

Applications

A titanium cylinder, "Grade 2" quality

Titanium is used in steel as an alloying element (ferro-titanium) to reduce grain size and as a de-oxidizer, and in stainless steel to reduce carbon content. Titanium is often alloyed with aluminium (to refine grain size), vanadium, copper (to harden), iron, manganese, molybdenum, and other metals. Titanium mill products (sheet, plate, bar, wire, forgings, castings) find application in industrial, aerospace, recreational, and emerging markets. Powdered titanium is used in pyrotechnics as a source of bright-burning particles.

Pigments, Additives and Coatings

Titanium dioxide is the most commonly used compound of titanium

About 95% of all titanium ore is destined for refinement into titanium dioxide (TiO_2), an intensely white permanent pigment used in paints, paper, toothpaste, and plastics. It is also used in cement, in gemstones, as an optical opacifier in paper, and a strengthening agent in graphite composite fishing rods and golf clubs.

TiO_2 powder is chemically inert, resists fading in sunlight, and is very opaque: it imparts a pure and brilliant white color to the brown or gray chemicals that form the majority of household plastics. In nature, this compound is found in the minerals anatase, brookite, and rutile. Paint made with titanium dioxide does well in severe temperatures and marine environments. Pure titanium dioxide has a very high index of refraction and an optical dispersion higher than diamond. In addition to being a very important pigment, titanium dioxide is also used in sunscreens.

Aerospace and Marine

Because of their high tensile strength to density ratio, high corrosion resistance, fatigue resistance, high crack resistance, and ability to withstand moderately high temperatures without creeping, titanium alloys are used in aircraft, armor plating, naval ships, spacecraft, and missiles. For these applications, titanium is alloyed with aluminium, zirconium, nickel, vanadium, and other elements to manufacture a variety of components including critical structural parts, fire walls, landing gear, exhaust ducts (helicopters), and hydraulic systems. In fact, about two thirds of all titanium metal produced is used in aircraft engines and frames. The SR-71 "Blackbird" was one of the first aircraft frames where titanium was used, paving the way for much wider use in modern military and commercial aircraft. An estimated 59 metric tons (130,000 pounds) are used in the Boeing 777, 45 in the Boeing 747, 18 in the Boeing 737, 32 in the Airbus A340, 18 in the Airbus A330, and 12 in the Airbus A320. The Airbus A380 may use 77 metric tons, including about 11 tons in the engines. In engine applications, titanium is used for rotors, compressor blades, hydraulic system components, and nacelles. The titanium 6AL-4V alloy accounts for almost 50% of all alloys used in aircraft applications.

Because it is highly resistant to corrosion by sea water, titanium is used to make propeller shafts, rigging, and heat exchangers in desalination plants; heater-chillers for salt water aquariums, fishing line and leader, and divers' knives. Titanium is used in the housings and components of ocean-deployed surveillance and monitoring devices for science and the military. The former Soviet Union developed techniques for making submarines with hulls of titanium alloys forging titanium in huge vacuum tubes.

Industrial

High-purity (99.999%) titanium with visible crystallites

Welded titanium pipe and process equipment (heat exchangers, tanks, process vessels, valves) are used in the chemical and petrochemical industries primarily for corrosion resistance. Specific alloys are used in downhole and nickel hydrometallurgy for their high strength (e. g.: titanium Beta C alloy), corrosion resistance, or both. The pulp and paper industry uses titanium in process equipment exposed to corrosive media, such as sodium hypochlorite or wet chlorine gas (in the bleachery). Other applications include: ultrasonic welding, wave soldering, and sputtering targets.

Titanium tetrachloride ($TiCl_4$), a colorless liquid, is important as an intermediate in the process of making TiO_2 and is also used to produce the Ziegler–Natta catalyst. Titanium tetrachloride is also used to iridize glass and, because it fumes strongly in moist air, it is used to make smoke screens.

Consumer and Architectural

Titanium metal is used in automotive applications, particularly in automobile and motorcycle racing where low weight and high strength and rigidity are critical. The metal is generally too expensive for the general consumer market, though some late model Corvettes have been manufactured with titanium exhausts, and the new Corvette Z06's LT4 supercharged engine uses lightweight, solid titanium intake valves for greater strength and resistance to heat.

Titanium is used in many sporting goods: tennis rackets, golf clubs, lacrosse stick shafts; cricket, hockey, lacrosse, football helmet grills, and bicycle frames and components. Although not a mainstream material for bicycle production, titanium bikes have been used by racing teams and adventure cyclists.

Titanium alloys are used in spectacle frames that are rather expensive but highly durable, long lasting, light weight, and cause no skin allergies. Many backpackers use titanium equipment, including cookware, eating utensils, lanterns, and tent stakes. Though slightly more expensive than traditional steel or aluminium alternatives, titanium products can be significantly lighter without compromising strength. Titanium horseshoes are preferred to steel by farriers because it is lighter and more durable.

Titanium has occasionally been used in architecture. The 40 m (131 foot) memorial to Yuri Gagarin, the first man to travel in space, (55°42′29.7″N 37°34′57.2″E / 55.708250°N 37.582556°E / 55.708250; 37.582556), as well as the 110 m (360.9 feet) Monument to the Conquerors of Space on top of the Cosmonaut Museum in Moscow are made of titanium for the metal's attractive color and association with rocketry. The Guggenheim Museum Bilbao and the Cerritos Millennium Library were the first buildings in Europe and North America, respectively, to be sheathed in titanium panels. Titanium sheathing was used in the Frederic C. Hamilton Building in Denver, Colorado.

Because of its superior strength and light weight relative to other metals (steel, stainless steel, and aluminium), and because of recent advances in metalworking techniques, titanium has become more widespread in the manufacture of firearms. Primary uses include pistol frames and revolver cylinders. For the same reasons, it is used in the body of laptop computers (for example, in Apple's PowerBook line).

Some upmarket lightweight and corrosion-resistant tools, such as shovels and flashlights, are made of titanium or titanium alloys.

Jewelry

Because of its durability, titanium has become more popular for designer jewelry (particularly, titanium rings). Its inertness makes it a good choice for those with allergies or those who will be wearing the jewelry in environments such as swimming pools. Titanium is also alloyed with gold to produce an alloy that can be marketed as 24-carat gold because the 1% of alloyed Ti is insufficient to require a lesser mark. The resulting alloy is roughly the hardness of 14-carat gold and is more durable than pure 24-carat gold.

Titanium's durability, light weight, dent and corrosion resistance makes it useful for watch cases. Some artists work with titanium to produce sculptures, decorative objects and furniture.

Titanium may be anodized to vary the thickness of the surface oxide layer, causing optical interference fringes and a variety of bright colors. With this coloration and chemical inertness, titanium is a popular metal for body piercing.

Titanium has a minor use in dedicated non-circulating coins and medals. In 1999, Gibraltar released world's first titanium coin for the millennium celebration. The Gold Coast Titans, an Australian rugby league team, award a medal of pure titanium to their player of the year.

Medical

Because it is biocompatible (non-toxic and not rejected by the body), titanium has many medical uses, including surgical implements and implants, such as hip balls and sockets (joint replacement) and dental implants that can stay in place for up to 20 years. The titanium is often alloyed with about 4% aluminium or 6% Al and 4% vanadium.

Titanium has the inherent ability to osseointegrate, enabling use in dental implants that can last for over 30 years. This property is also useful for orthopedic implant applications. These benefit from titanium's lower modulus of elasticity (Young's modulus) to more closely match that of the bone that such devices are intended to repair. As a result, skeletal loads are more evenly shared between bone and implant, leading to a lower incidence of bone degradation due to stress shielding and periprosthetic bone fractures, which occur at the boundaries of orthopedic implants. However, titanium alloys' stiffness is still more than twice that of bone, so adjacent bone bears a greatly reduced load and may deteriorate.

Because titanium is non-ferromagnetic, patients with titanium implants can be safely examined with magnetic resonance imaging (convenient for long-term implants). Preparing titanium for implantation in the body involves subjecting it to a high-temperature plasma arc which removes the surface atoms, exposing fresh titanium that is instantly oxidized.

Titanium is also used for the surgical instruments used in image-guided surgery, as well as wheelchairs, crutches, and any other products where high strength and low weight are desirable.

Titanium dioxide nanoparticles are widely used in electronics and the delivery of pharmaceuticals and cosmetics.

Nuclear Waste Storage

Because of its excellent corrosion resistance, titanium containers have been studied for the long-term storage of nuclear waste. Containers lasting more than 100,000 years are possible with manufacturing conditions that minimize material defects. A titanium "drip shield" could also be installed over containers of other types to enhance their longevity.

Bioremediation

The fungal species *Marasmius oreades* and *Hypholoma capnoides* can bio convert titanium in titanium polluted soils.

Precautions

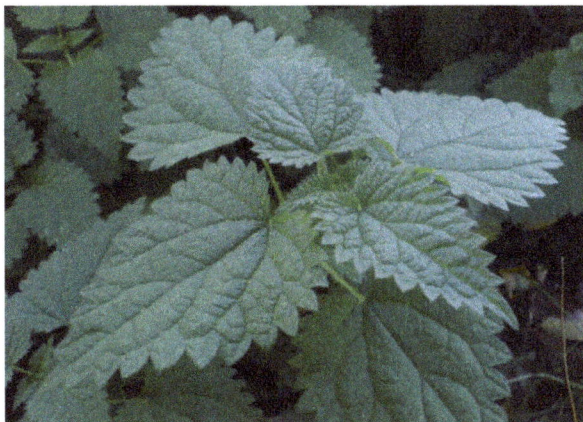

Nettles contain up to 80 parts per million of titanium.

Titanium is non-toxic even in large doses and does not play any natural role inside the human body. An estimated quantity of 0.8 milligrams of titanium is ingested by humans each day, but most passes through without being absorbed in the tissues. It does, however, sometimes bio-accumulate in tissues that contain silica. One study indicates a possible connection between titanium and yellow nail syndrome. An unknown mechanism in plants may use titanium to stimulate the production of carbohydrates and encourage growth. This may explain why most plants contain about 1 part per million (ppm) of titanium, food plants have about 2 ppm, and horsetail and nettle contain up to 80 ppm.

As a powder or in the form of metal shavings, titanium metal poses a significant fire hazard and, when heated in air, an explosion hazard. Water and carbon dioxide are ineffective for extinguishing a titanium fire; Class D dry powder agents must be used instead.

When used in the production or handling of chlorine, titanium should not be exposed to dry chlorine gas because it may result in a titanium/chlorine fire. Even wet chlorine presents a fire hazard when extreme weather conditions cause unexpected drying.

Titanium can catch fire when a fresh, non-oxidized surface comes in contact with liquid oxygen. Fresh metal may be exposed when the oxidized surface is struck or scratched with a hard object, or when mechanical strain causes a crack. This poses a limitation to its use in liquid oxygen systems, such as those in the aerospace industry. Because titanium tubing impurities can cause fires when exposed to oxygen, titanium is prohibited in gaseous oxygen respiration systems. Steel tubing is used for high pressure systems (3,000 p.s.i.) and aluminium tubing for low pressure systems.

Beryllium

Beryllium is a chemical element with symbol Be and atomic number 4. It is a relatively rare element in the universe, usually occurring as a product of the spallation of larger atomic nuclei that have collided with cosmic rays. Within the cores of stars beryllium is depleted as it is fused and creates larger elements. It is a divalent element which occurs naturally only in combination with other elements in minerals. Notable gemstones which contain beryllium include beryl (aquamarine, emerald) and chrysoberyl. As a free element it is a steel-gray, strong, lightweight and brittle alkaline earth metal.

Beryllium improves many physical properties when added as an alloying element to aluminium, copper (notably the alloy beryllium copper), iron and nickel. Beryllium does not form oxides until it reaches very high temperatures. Tools made of beryllium copper alloys are strong and hard and do not create sparks when they strike a steel surface. In structural applications, the combination of high flexural rigidity, thermal stability, thermal conductivity and low density (1.85 times that of water) make beryllium metal a desirable aerospace material for aircraft components, missiles, spacecraft, and satellites. Because of its low density and atomic mass, beryllium is relatively transparent to X-rays and other forms of ionizing radiation; therefore, it is the most common window material for X-ray equipment and components of particle physics experiments. The high thermal conductivities of beryllium and beryllium oxide have led to their use in thermal management applications.

The commercial use of beryllium requires the use of appropriate dust control equipment and industrial controls at all times because of the toxicity of inhaled beryllium-containing dusts that can cause a chronic life-threatening allergic disease in some people called berylliosis.

Characteristics

Physical Properties

Beryllium is a steel gray and hard metal that is brittle at room temperature and has a close-packed hexagonal crystal structure. It has exceptional stiffness (Young's modulus 287 GPa) and a reasonably high melting point. The modulus of elasticity of beryllium is approximately 50% greater than that of steel. The combination of this modulus and a relatively low density results in an unusually fast sound conduction speed in beryllium – about 12.9 km/s at ambient conditions. Other significant properties are high specific heat (1925 J·kg^{-1}·K^{-1}) and thermal conductivity (216 W·m^{-1}·K^{-1}), which make beryllium the metal with the best heat dissipation characteristics per unit weight. In combination with the relatively low coefficient of linear thermal expansion (11.4×10^{-6} K^{-1}), these characteristics result in a unique stability under conditions of thermal loading.

Nuclear Properties

Naturally occurring beryllium, save for slight contamination by cosmogenic radioisotopes, is essentially pure beryllium-9, which has a nuclear spin of 3/2. Beryllium has a large scattering cross section for high-energy neutrons, about 6 barns for energies above approximately 10 keV. Therefore, it works as a neutron reflector and neutron moderator, effectively slowing the neutrons to the thermal energy range of below 0.03 eV, where the total cross section is at least an order of magnitude lower – exact value strongly depends on the purity and size of the crystallites in the material.

The single primordial beryllium isotope ^{9}Be also undergoes a (n,2n) neutron reaction with neutron energies over about 1.9 MeV, to produce ^{8}Be, which almost immediately breaks into two alpha particles. Thus, for high-energy neutrons, beryllium is a neutron multiplier, releasing more neutrons than it absorbs. This nuclear reaction is:

$$^{9}_{4}\text{Be} + n \rightarrow 2(^{4}_{2}\text{He}) + 2n$$

Neutrons are liberated when beryllium nuclei are struck by energetic alpha particles producing the nuclear reaction

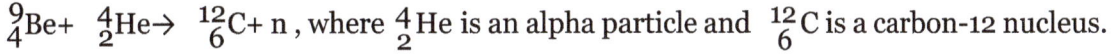

$$^{9}_{4}\text{Be} + \; ^{4}_{2}\text{He} \rightarrow \; ^{12}_{6}\text{C} + n \text{ , where } ^{4}_{2}\text{He is an alpha particle and } ^{12}_{6}\text{C is a carbon-12 nucleus.}$$

Beryllium also releases neutrons under bombardment by gamma rays. Thus, natural beryllium bombarded either by alphas or gammas from a suitable radioisotope is a key component of most radioisotope-powered nuclear reaction neutron sources for the laboratory production of free neutrons.

Small amounts of tritium are liberated when $^{9}_{4}\text{Be}$ nuclei absorb low energy neutrons in the three-step nuclear reaction

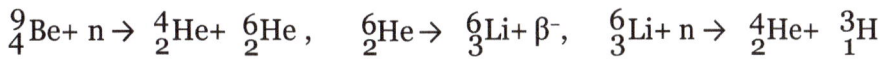

$$^{9}_{4}\text{Be} + n \rightarrow \; ^{4}_{2}\text{He} + \; ^{6}_{2}\text{He} , \quad ^{6}_{2}\text{He} \rightarrow \; ^{6}_{3}\text{Li} + \beta^{-}, \quad ^{6}_{3}\text{Li} + n \rightarrow \; ^{4}_{2}\text{He} + \; ^{3}_{1}\text{H}$$

Note that $^{6}_{2}\text{He}$ has a half life of only 0.8 seconds, β^{-} is an electron, and $^{6}_{3}\text{Li}$ has a high neutron absorption cross-section. Tritium is a radioisotope of concern in nuclear reactor waste streams.

As a metal, beryllium is transparent to most wavelengths of X-rays and gamma rays, making it useful for the output windows of X-ray tubes and other such apparatus.

Isotopes and Nucleosynthesis

Both stable and unstable isotopes of beryllium are created in stars, but the radioisotopes do not last long. It is believed that most of the stable beryllium in the universe was originally created in the interstellar medium when cosmic rays induced fission in heavier elements found in interstellar gas and dust. Primordial beryllium contains only one stable isotope, ^{9}Be, and therefore beryllium is a monoisotopic element.

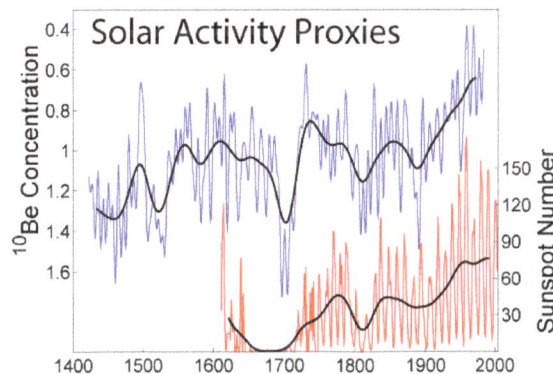

Plot showing variations in solar activity, including variation in sunspot number (red) and ^{10}Be concentration (blue). Note that the beryllium scale is inverted, so increases on this scale indicate lower ^{10}Be levels

Radioactive cosmogenic ^{10}Be is produced in the atmosphere of the Earth by the cosmic ray spallation of oxygen. ^{10}Be accumulates at the soil surface, where its relatively long half-life (1.36 million years) permits a long residence time before decaying to boron-10. Thus, ^{10}Be and its daughter products are used to examine natural soil erosion, soil formation and the development of lateritic soils, and as a proxy for measurement of the variations in solar activity and the age of ice cores. The production of ^{10}Be is inversely proportional to solar activity, because increased solar wind during periods of high solar activity decreases the flux of galactic cosmic rays that reach the Earth. Nuclear explosions also form ^{10}Be by the reaction of fast neutrons with ^{13}C in the carbon dioxide in air. This is one of the indicators of past activity at nuclear weapon test sites. The isotope ^{7}Be (half-life 53 days) is also cosmogenic, and shows an atmospheric abundance linked to sunspots, much like ^{10}Be.

^{8}Be has a very short half-life of about 7×10^{-17} s that contributes to its significant cosmological role, as elements heavier than beryllium could not have been produced by nuclear fusion in the Big Bang. This is due to the lack of sufficient time during the Big Bang's nucleosynthesis phase to produce carbon by the fusion of ^{4}He nuclei and the very low concentrations of available beryllium-8. The British astronomer Sir Fred Hoyle first showed that the energy levels of ^{8}Be and ^{12}C allow carbon production by the so-called triple-alpha process in helium-fueled stars where more nucleosynthesis time is available. This process allows carbon to be produced in stars, but not in the Big Bang. Star-created carbon (the basis of carbon-based life) is thus a component in the elements in the gas and dust ejected by AGB stars and supernovae, as well as the creation of all other elements with atomic numbers larger than that of carbon.

The 2s electrons of beryllium may contribute to chemical bonding. Therefore, when ^{7}Be decays by L-electron capture, it does so by taking electrons from its atomic orbitals that may be participating in bonding. This makes its decay rate dependent to a measurable degree upon its chemical surroundings – a rare occurrence in nuclear decay.

The shortest-lived known isotope of beryllium is ^{13}Be which decays through neutron emission. It has a half-life of 2.7×10^{-21} s. ^{6}Be is also very short-lived with a half-life of 5.0×10^{-21} s. The exotic isotopes ^{11}Be and ^{14}Be are known to exhibit a nuclear halo. This phenomenon can be understood as the nuclei of ^{11}Be and ^{14}Be have, respectively, 1 and 4 neutrons orbiting substantially outside the classical Fermi 'waterdrop' model of the nucleus.

Occurrence

Beryllium ore with 1US¢ coin for scale

Emerald is a naturally occurring compound of beryllium.

The Sun has a concentration of 0.1 parts per billion (ppb) of beryllium. Beryllium has a concentration of 2 to 6 parts per million (ppm) in the Earth's crust. It is most concentrated in the soils, 6 ppm. Trace amounts of ^{9}Be are found in the Earth's atmosphere. The concentration of beryllium in sea water is 0.2–0.6 parts per trillion. In stream water, however, beryllium is more abundant with a concentration of 0.1 ppb.

Beryllium is found in over 100 minerals, but most are uncommon to rare. The more common beryllium containing minerals include: bertrandite ($Be_4Si_2O_7(OH)_2$), beryl ($Al_2Be_3Si_6O_{18}$), chrysoberyl (Al_2BeO_4) and phenakite (Be_2SiO_4). Precious forms of beryl are aquamarine, red beryl and emerald. The green color in gem-quality forms of beryl comes from varying amounts of chromium (about 2% for emerald).

The two main ores of beryllium, beryl and bertrandite, are found in Argentina, Brazil, India, Madagascar, Russia and the United States. Total world reserves of beryllium ore are greater than 400,000 tonnes.

Production

The extraction of beryllium from its compounds is a difficult process due to its high affinity for oxygen at elevated temperatures, and its ability to reduce water when its oxide film is removed. The United States, China and Kazakhstan are the only three countries involved in the industrial-scale extraction of beryllium.

Beryllium is most commonly extracted from the mineral beryl, which is either sintered using an extraction agent or melted into a soluble mixture. The sintering process involves mixing beryl with sodium fluorosilicate and soda at 770 °C (1,420 °F) to form sodium fluoroberyllate, aluminium oxide and silicon dioxide. Beryllium hydroxide is precipitated from a solution of sodium fluoroberyllate and sodium hydroxide in water. Extraction of beryllium using the melt method involves grinding beryl into a powder and heating it to 1,650 °C (3,000 °F). The melt is quickly cooled with water and then reheated 250 to 300 °C (482 to 572 °F) in concentrated sulfuric acid, mostly yielding beryllium sulfate and aluminium sulfate. Aqueous ammonia is then used to remove the aluminium and sulfur, leaving beryllium hydroxide.

Beryllium hydroxide created using either the sinter or melt method is then converted into beryllium fluoride or beryllium chloride. To form the fluoride, aqueous ammonium hydrogen fluoride is added to beryllium hydroxide to yield a precipitate of ammonium tetrafluoroberyllate, which is heated to 1,000 °C (1,830 °F) to form beryllium fluoride. Heating the fluoride to 900 °C (1,650 °F) with magnesium forms finely divided beryllium, and additional heating to 1,300 °C (2,370 °F) creates the compact metal. Heating beryllium hydroxide forms the oxide, which becomes beryllium chloride when combined with carbon and chlorine. Electrolysis of molten beryllium chloride is then used to obtain the metal.

Chemical Properties

Beryllium's chemical behavior is largely a result of its small atomic and ionic radii. It thus has very high ionization potentials and strong polarization while bonded to other atoms, which is why all of its compounds are covalent. It is more chemically similar to aluminium than its close neighbors in the periodic table due to having a similar charge-to-radius ratio. An oxide layer forms around beryllium that prevents further reactions with air unless heated above 1000 °C. Once ignited, beryllium burns brilliantly forming a mixture of beryllium oxide and beryllium nitride. Beryllium dissolves readily in non-oxidizing acids, such as HCl and diluted H_2SO_4, but not in nitric acid or water as this forms the oxide. This behavior is similar to that of aluminium metal. Beryllium also dissolves in alkali solutions.

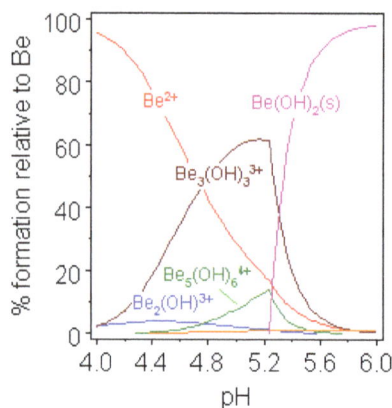

Beryllium hydrolysis as a function of pH Water molecules attached to Be are omitted

The beryllium atom has the electronic configuration [He] $2s^2$. The two valence electrons give beryllium a +2 oxidation state and thus the ability to form two covalent bonds; the only evidence of lower valence of beryllium is in the solubility of the metal in $BeCl_2$. Due to the octet rule, atoms tend to seek a valence of 8 in order to resemble a noble gas. Beryllium tries to achieve a coordination number of 4 because its two covalent bonds fill half of this octet. Tetracoordination allows beryllium compounds, such as the fluoride or chloride, to form polymers.

This characteristic is employed in analytical techniques using EDTA as a ligand. EDTA preferentially forms octahedral complexes – thus absorbing other cations such as Al^{3+} which might interfere – for example, in the solvent extraction of a complex formed between Be^{2+} and acetylacetone. Beryllium(II) readily forms complexes with strong donating ligands such as phosphine oxides and arsine oxides. There have been extensive studies of these complexes which show the stability of the O-Be bond.

Solutions of beryllium salts, e.g. beryllium sulfate and beryllium nitrate, are acidic because of hydrolysis of the $[Be(H_2O)_4]^{2+}$ ion.

$$[Be(H_2O)_4]^{2+} + H_2O \rightleftharpoons [Be(H_2O)_3(OH)]^+ + H_3O^+$$

Other products of hydrolysis include the trimeric ion $[Be_3(OH)_3(H_2O)_6]^{3+}$. Beryllium hydroxide, $Be(OH)_2$, is insoluble even in acidic solutions with pH less than 6, that is at biological pH. It is amphoteric and dissolves in strongly alkaline solutions.

Beryllium forms binary compounds with many non-metals. Anhydrous halides are known for F, Cl, Br and I. BeF_2 has a silica-like structure with corner-shared BeF_4 tetrahedra. $BeCl_2$ and $BeBr_2$ have chain structures with edge-shared tetrahedra. All beryllium halides have a linear monomeric molecular structure in the gas phase.

Beryllium difluoride, BeF_2, is different than the other difluorides. In general, beryllium has a tendency to bond covalently, much more so than the other alkaline earths and its fluoride is partially covalent (although still more ionic than its other halides). BeF_2 has many similarities to SiO_2 (quartz) a mostly covalently bonded network solid. BeF_2 has tetrahedrally coordinated metal and forms glasses (is difficult to crystallize). When crystalline, beryllium fluoride has the same room temperature crystal structure as quartz and shares many higher temperature structures also. Beryllium difluoride is very soluble in water, unlike the other alkaline earth difluorides. (Although they are strongly ionic, they do not dissolve because of the especially strong lattice energy of the fluorite structure.) However, BeF_2 has much lower electrical conductivity when in solution or when molten than would be expected if it were fully ionic.

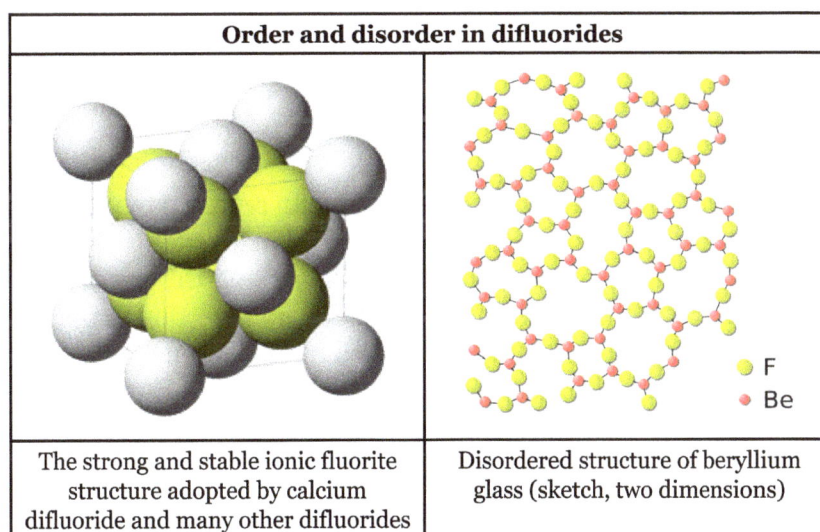

Order and disorder in difluorides

The strong and stable ionic fluorite structure adopted by calcium difluoride and many other difluorides	Disordered structure of beryllium glass (sketch, two dimensions)

Beryllium oxide, BeO, is a white refractory solid, which has the wurtzite crystal structure and a thermal conductivity as high as in some metals. BeO is amphoteric. Salts of beryllium can be produced by treating $Be(OH)_2$ with acid. Beryllium sulfide, selenide and telluride are known, all having the zincblende structure.

Beryllium nitride, Be_3N_2 is a high-melting-point compound which is readily hydrolyzed. Beryllium azide, BeN_6 is known and beryllium phosphide, Be_3P_2 has a similar structure to Be_3N_2. Basic beryllium nitrate and basic beryllium acetate have similar tetrahedral structures with four beryllium atoms coordinated to a central oxide ion. A number of beryllium borides are known, such as Be_5B, Be_4B, Be_2B, BeB_2, BeB_6 and BeB_{12}. Beryllium carbide, Be_2C, is a refractory brick-red compound that reacts with water to give methane. No beryllium silicide has been identified.

History

The mineral beryl, which contains beryllium, has been used at least since the Ptolemaic dynasty of Egypt. In the first century CE, Roman naturalist Pliny the Elder mentioned in his encyclopedia *Natural History* that beryl and emerald ("smaragdus") were similar. The Papyrus Graecus Holmiensis, written in the third or fourth century CE, contains notes on how to prepare artificial emerald and beryl.

Louis-Nicolas Vauquelin discovered beryllium

Early analyses of emeralds and beryls by Martin Heinrich Klaproth, Torbern Olof Bergman, Franz Karl Achard, and Johann Jakob Bindheim always yielded similar elements, leading to the fallacious conclusion that both substances are aluminium silicates. Mineralogist René Just Haüy discovered that both crystals are geometrically identical, and he asked chemist Louis-Nicolas Vauquelin for a chemical analysis.

In a 1798 paper read before the Institut de France, Vauquelin reported that he found a new "earth" by dissolving aluminium hydroxide from emerald and beryl in an additional alkali. The editors of the journal *Annales de Chimie et de Physique* named the new earth "glucine" for the sweet taste of some of its compounds. Klaproth preferred the name "beryllina" due to the fact that yttria also formed sweet salts. The name "beryllium" was first used by Wöhler in 1828.

Friedrich Wöhler was one of the men who independently isolated beryllium

Friedrich Wöhler and Antoine Bussy independently isolated beryllium in 1828 by the chemical reaction of metallic potassium with beryllium chloride, as follows:

$$BeCl_2 + 2\,K \rightarrow 2\,KCl + Be$$

Using an alcohol lamp, Wöhler heated alternating layers of beryllium chloride and potassium in a wired-shut platinum crucible. The above reaction immediately took place and caused the crucible to become white hot. Upon cooling and washing the resulting gray-black powder he saw that it was made of fine particles with a dark metallic luster. The highly reactive potassium had been produced by the electrolysis of its compounds, a process discovered 21 years before. The chemical method using potassium yielded only small grains of beryllium from which no ingot of metal could be cast or hammered.

The direct electrolysis of a molten mixture of beryllium fluoride and sodium fluoride by Paul Lebeau in 1898 resulted in the first pure (99.5 to 99.8%) samples of beryllium. The first commercially successful process for producing beryllium was developed in 1932 by Alfred Stock and Hans Goldschmidt. Their process involves the electrolysis of a mixture of beryllium fluorides and barium, which causes molten beryllium to collect on a water-cooled iron cathode.

A sample of beryllium was bombarded with alpha rays from the decay of radium in a 1932 experiment by James Chadwick that uncovered the existence of the neutron. This same method is used in one class of radioisotope-based laboratory neutron sources that produce 30 neutrons for every million α particles.

Beryllium production saw a rapid increase during World War II, due to the rising demand for hard beryllium-copper alloys and phosphors for fluorescent lights. Most early fluorescent lamps used zinc orthosilicate with varying content of beryllium to emit greenish light. Small additions of magnesium tungstate improved the blue part of the spectrum to yield an acceptable white light. Halophosphate-based phosphors replaced beryllium-based phosphors after beryllium was found to be toxic.

Electrolysis of a mixture of beryllium fluoride and sodium fluoride was used to isolate beryllium during the 19th century. The metal's high melting point makes this process more energy-consuming than corresponding processes used for the alkali metals. Early in the 20th century, the production of beryllium by the thermal decomposition of beryllium iodide was investigated following the success of a similar process for the production of zirconium, but this process proved to be uneconomical for volume production.

Pure beryllium metal did not become readily available until 1957, even though it had been used as an alloying metal to harden and toughen copper much earlier. Beryllium could be produced by reducing beryllium compounds such as beryllium chloride with metallic potassium or sodium. Currently most beryllium is produced by reducing beryllium fluoride with purified magnesium. The price on the American market for vacuum-cast beryllium ingots was about $338 per pound ($745 per kilogram) in 2001.

Between 1998 and 2008, the world's production of beryllium had decreased from 343 to about 200 tonnes, of which 176 tonnes (88%) came from the United States.

Applications

Radiation Windows

Beryllium target which "converts" a proton beam into a neutron beam

A square beryllium foil mounted in a steel case to be used as a window between a vacuum chamber and an X-ray microscope. Beryllium is highly transparent to X-rays owing to its low atomic number.

Because of its low atomic number and very low absorption for X-rays, the oldest and still one of the most important applications of beryllium is in radiation windows for X-ray tubes. Extreme demands are placed on purity and cleanliness of beryllium to avoid artifacts in the X-ray images. Thin beryllium foils are used as radiation windows for X-ray detectors, and the extremely low absorption minimizes the heating effects caused by high intensity, low energy X-rays typical of synchrotron radiation. Vacuum-tight windows and beam-tubes for radiation experiments on synchrotrons are manufactured exclusively from beryllium. In scientific setups for various X-ray emission studies (e.g., energy-dispersive X-ray spectroscopy) the sample holder is usually made of beryllium because its emitted X-rays have much lower energies (~100 eV) than X-rays from most studied materials.

Low atomic number also makes beryllium relatively transparent to energetic particles. Therefore, it is used to build the beam pipe around the collision region in particle physics setups, such as all four main detector experiments at the Large Hadron Collider (ALICE, ATLAS, CMS, LHCb), the Tevatron and the SLAC. The low density of beryllium allows collision products to reach the surrounding detectors without significant interaction, its stiffness allows a powerful vacuum to be produced within the pipe to minimize interaction with gases, its thermal stability allows it to function correctly at temperatures of only a few degrees above absolute zero, and its diamagnetic nature keeps it from interfering with the complex multipole magnet systems used to steer and focus the particle beams.

Mechanical Applications

Because of its stiffness, light weight and dimensional stability over a wide temperature range, beryllium metal is used for lightweight structural components in the defense and aerospace industries in high-speed aircraft, guided missiles, spacecraft, and satellites. Several liquid-fuel rockets have used rocket nozzles made of pure beryllium. Beryllium powder was itself studied as a rocket fuel, but this use has never materialized. A small number of extreme high-end bicycle frames have been built with beryllium. From 1998 to 2000, the McLaren Formula One team used Mercedes-Benz engines with beryllium-aluminium-alloy pistons. The use of beryllium engine components was banned following a protest by Scuderia Ferrari.

Mixing about 2.0% beryllium into copper forms an alloy called beryllium copper that is six times stronger than copper alone. Beryllium alloys are used in many applications because of their combination of elasticity, high electrical conductivity and thermal conductivity, high strength and hardness, nonmagnetic properties, as well as good corrosion and fatigue resistance. These applications include non-sparking tools that are used near flammable gases (beryllium nickel), in springs and membranes (beryllium nickel and beryllium iron) used in surgical instruments and high temperature devices. As little as 50 parts per million of beryllium alloyed with liquid magnesium leads to a significant increase in oxidation resistance and decrease in flammability.

Beryllium Copper Adjustable Wrench

The high elastic stiffness of beryllium has led to its extensive use in precision instrumentation, e.g. in inertial guidance systems and in the support mechanisms for optical systems. Beryllium-copper alloys were also applied as a hardening agent in "Jason pistols", which were used to strip the paint from the hulls of ships.

Beryllium was also used for cantilevers in high performance phonograph cartridge styli, where its extreme stiffness and low density allowed for tracking weights to be reduced to 1 gram, yet still track high frequency passages with minimal distortion.

An earlier major application of beryllium was in brakes for military airplanes because of its hardness, high melting point, and exceptional ability to dissipate heat. Environmental considerations have led to substitution by other materials.

To reduce costs, beryllium can be alloyed with significant amounts of aluminium, resulting in the AlBeMet alloy (a trade name). This blend is cheaper than pure beryllium, while still retaining many desirable properties.

Mirrors

Beryllium mirrors are of particular interest. Large-area mirrors, frequently with a honeycomb support structure, are used, for example, in meteorological satellites where low weight and long-term dimensional stability are critical. Smaller beryllium mirrors are used in optical guidance systems and in fire-control systems, e.g. in the German-made Leopard 1 and Leopard 2 main battle tanks. In these systems, very rapid movement of the mirror is required which again dictates low mass and high rigidity. Usually the beryllium mirror is coated with hard electroless nickel plating which can be more easily polished to a finer optical finish than beryllium. In some applications, though, the beryllium blank is polished without any coating. This is particularly applicable to cryogenic operation where thermal expansion mismatch can cause the coating to buckle.

The James Webb Space Telescope will have 18 hexagonal beryllium sections for its mirrors. Because JWST will face a temperature of 33 K, the mirror is made of gold-plated beryllium, capable of handling extreme cold better than glass. Beryllium contracts and deforms less than glass – and remains more uniform – in such temperatures. For the same reason, the optics of the Spitzer Space Telescope are entirely built of beryllium metal.

Magnetic Applications

Sphere Beryllium B52 - Gyrocompass

Beryllium is non-magnetic. Therefore, tools fabricated out of beryllium-based materials are used by naval or military explosive ordnance disposal teams for work on or near naval mines, since these mines commonly have magnetic fuzes. They are also found in maintenance and construction materials near magnetic resonance imaging (MRI) machines because of the high magnetic fields generated. In the fields of radio communications and powerful (usually military) radars, hand tools made of beryllium are used to tune the highly magnetic klystrons, magnetrons, traveling wave tubes, etc., that are used for generating high levels of microwave power in the transmitters.

Nuclear Applications

Thin plates or foils of beryllium are sometimes used in nuclear weapon designs as the very outer layer of the plutonium pits in the primary stages of thermonuclear bombs, placed to surround the fissile material. These layers of beryllium are good "pushers" for the implosion of the plutonium-239, and they are also good neutron reflectors, just as they are in beryllium-moderated nuclear reactors.

Beryllium is also commonly used in some neutron sources in laboratory devices in which relatively few neutrons are needed (rather than having to use a nuclear reactor, or a particle accelerator-powered neutron generator). For this purpose, a target of beryllium-9 is bombarded with energetic alpha particles from a radioisotope such as polonium-210, radium-226, plutonium-239, or americium-241. In the nuclear reaction that occurs, a beryllium nucleus is transmuted into carbon-12, and one free neutron is emitted, traveling in about the same direction as the alpha particle was heading. Such alpha decay driven beryllium neutron sources, named "urchin" neutron initiators, were used some in early atomic bombs. Neutron sources in which beryllium is bombarded with gamma rays from a gamma decay radioisotope, are also used to produce laboratory neutrons.

Two CANDU fuel bundles: Each about 50 cm in length and 10 cm in diameter. Notice the small appendages on the fuel clad surfaces

Beryllium is also used in fuel fabrication for CANDU reactors. The fuel elements have small appendages that are resistance brazed to the fuel cladding using an induction brazing process with Be as the braze filler material. Bearing pads are brazed on to prevent fuel bundle to pressure tube contact, and inter-element spacer pads are brazed on to prevent element to element contact.

Beryllium is also used at the Joint European Torus nuclear-fusion research laboratory, and it will be used in the more advanced ITER to condition the components which face the plasma. Beryllium has also been proposed as a cladding material for nuclear fuel rods, because of its good combination of mechanical, chemical, and nuclear properties. Beryllium fluoride is one of the constituent salts of the eutectic salt mixture FLiBe, which is used as a solvent, moderator and coolant in many hypothetical molten salt reactor designs, including the liquid fluoride thorium reactor (LFTR).

Acoustics

The low weight and high rigidity of beryllium make it useful as a material for high-frequency speaker drivers. Because beryllium is expensive (many times more than titanium), hard to shape due to its brittleness, and toxic if mishandled, beryllium tweeters are limited to high-end home, pro audio, and public address applications. Some high-fidelity products have been fraudulently claimed to be made of the material.

Electronic

Beryllium is a p-type dopant in III-V compound semiconductors. It is widely used in materials such as GaAs, AlGaAs, InGaAs and InAlAs grown by molecular beam epitaxy (MBE). Cross-rolled beryllium sheet is an excellent structural support for printed circuit boards in surface-mount technology. In critical electronic applications, beryllium is both a structural support and heat sink. The application also requires a coefficient of thermal expansion that is well matched to the alumina and polyimide-glass substrates. The beryllium-beryllium oxide composite "E-Materials" have been specially designed for these electronic applications and have the additional advantage that the thermal expansion coefficient can be tailored to match diverse substrate materials.

Beryllium oxide is useful for many applications that require the combined properties of an electrical insulator and an excellent heat conductor, with high strength and hardness, and a very high melting point. Beryllium oxide is frequently used as an insulator base plate in high-power transistors in radio frequency transmitters for telecommunications. Beryllium oxide is also being studied for use in increasing the thermal conductivity of uranium dioxide nuclear fuel pellets. Beryllium compounds were used in fluorescent lighting tubes, but this use was discontinued because of the disease berylliosis which developed in the workers who were making the tubes.

Healthcare

Beryllium is a component of several dental alloys.

Precautions

Approximately 35 micrograms of beryllium is found in the average human body, an amount not considered harmful. Beryllium is chemically similar to magnesium and therefore can displace it from enzymes, which causes them to malfunction. Because Be^{2+} is a highly charged and small ion, it can easily get into many tissues and cells, where it specifically targets cell nuclei, inhibiting many enzymes, including those used for synthesizing DNA. Its toxicity is exacerbated by the fact that the body has no means to control beryllium levels, and once inside the body the beryllium cannot be removed. Chronic berylliosis is a pulmonary and systemic granulomatous disease caused by inhalation of dust or fumes contaminated with beryllium; either large amounts over a short time or small amounts over a long time can lead to this ailment. Symptoms of the disease can take up to five years to develop; about a third of patients with it die and the survivors are left disabled. The International Agency for Research on Cancer (IARC) lists beryllium and beryllium compounds as Category 1 carcinogens. In the US, the Occupational Safety and Health Administration (OSHA) has designated a permissible exposure limit (PEL) in the workplace with a time-weighted average (TWA) 0.002 mg/m³ and a constant exposure limit of 0.005 mg/m³ over 30 minutes, with a maximum peak limit of 0.025 mg/m³. The National Institute for Occupational Safety and Health (NIOSH) has set a recommended exposure limit (REL) of constant 0.0005 mg/m³. The IDLH (immediately dangerous to life and health) value is 4 mg/m³.

The toxicity of finely divided beryllium (dust or powder, mainly encountered in industrial settings where beryllium is produced or machined) is very well-documented. Solid beryllium metal does not carry the same hazards as airborne inhaled dust, but any hazard associated with physical contact is poorly documented. Workers handling finished beryllium pieces are routinely advised to handle them with gloves, both as a precaution and because many if not most applications of beryllium cannot tolerate residue of skin contact such as fingerprints.

Acute beryllium disease in the form of chemical pneumonitis was first reported in Europe in 1933 and in the United States in 1943. A survey found that about 5% of workers in plants manufacturing fluorescent lamps in 1949 in the United States had beryllium-related lung diseases. Chronic berylliosis resembles sarcoidosis in many respects, and the differential diagnosis is often difficult. It killed some early workers in nuclear weapons design, such as Herbert L. Anderson.

Beryllium may be found in coal slag. When the slag is formulated into an abrasive agent for blasting paint and rust from hard surfaces, the beryllium can become airborne and become a source of exposure.

Early researchers tasted beryllium and its various compounds for sweetness in order to verify its presence. Modern diagnostic equipment no longer necessitates this highly risky procedure and no attempt should be made to ingest this highly toxic substance. Beryllium and its compounds should be handled with great care and special precautions must be taken when carrying out any activity which could result in the release of beryllium dust (lung cancer is a possible result of prolonged exposure to beryllium-laden dust). Although the use of beryllium compounds in fluorescent lighting tubes was discontinued in 1949, potential for exposure to beryllium exists in the nuclear and aerospace industries and in the refining of beryllium metal and melting of beryllium-containing alloys, the manufacturing of electronic devices, and the handling of other beryllium-containing material.

A successful test for beryllium in air and on surfaces has been recently developed and published as an international voluntary consensus standard ASTM D7202. The procedure uses dilute ammonium bifluoride for dissolution and fluorescence detection with beryllium bound to sulfonated hydroxybenzoquinoline, allowing up to 100 times more sensitive detection than the recommended limit for beryllium concentration in the workplace. Fluorescence increases with increasing beryllium concentration. The new procedure has been successfully tested on a variety of surfaces and is effective for the dissolution and ultratrace detection of refractory beryllium oxide and siliceous beryllium (ASTM D7458).

Zirconium

Zirconium is a chemical element with symbol Zr and atomic number 40. The name of zirconium is taken from the name of the mineral zircon, the most important source of zirconium. The word *zircon* comes from the Persian word *zargun* زرگون, meaning "gold-colored". It is a lustrous, grey-white, strong transition metal that resembles hafnium and, to a lesser extent, titanium. Zirconium is mainly used as a refractory and opacifier, although small amounts are used as an alloying agent for its strong resistance to corrosion. Zirconium forms a variety of inorganic and organometallic compounds such as zirconium dioxide and zirconocene dichloride, respectively. Five isotopes occur naturally, three of which are stable. Zirconium compounds have no known biological role.

Characteristics

Zirconium is a lustrous, greyish-white, soft, ductile and malleable metal that is solid at room temperature, though it is hard and brittle at lesser purities. In powder form, zirconium is highly flammable, but the solid form is much less prone to ignition. Zirconium is highly resistant to corrosion by alkalis, acids, salt water and other agents. However, it will dissolve in hydrochloric and sulfuric acid, especially when fluorine is present. Alloys with zinc are magnetic at less than 35 K.

The melting point of zirconium is 1855 °C (3371 °F), and the boiling point is 4371 °C (7900 °F). Zirconium has an electronegativity of 1.33 on the Pauling scale. Of the elements within the d-block, zirconium has the fourth lowest electronegativity after yttrium, lanthanum (or lutetium), and hafnium.

At room temperature zirconium exhibits a hexagonally close packed crystal structure, α-Zr, which changes to β-Zr a body-centered cubic crystal structure at 863 °C. Zirconium exists in the β-phase until the melting point.

Isotopes

A zirconium rod

Naturally occurring zirconium is composed of five isotopes. ^{90}Zr, ^{91}Zr, ^{92}Zr and ^{94}Zr are stable, although ^{94}Zr is predicted to undergo double beta decay (not observed experimentally) with a half-life of more than 1.10×10^{17} years. ^{96}Zr has a half-life of 2.4×10^{19} years, and is the longest-lived radioisotope of zirconium. Of these natural isotopes, ^{90}Zr is the most common, making up 51.45% of all zirconium. ^{96}Zr is the least common, comprising only 2.80% of zirconium.

Twenty-eight artificial isotopes of zirconium have been synthesized, ranging in atomic mass from 78 to 110. ^{93}Zr is the longest-lived artificial isotope, with a half-life of 1.53×10^{6} years. ^{110}Zr, the heaviest isotope of zirconium, is the most radioactive, with an estimated half-life of 30 milliseconds. Radioactive isotopes at or above mass number 93 decay by electron emission, whereas those at or below 89 decay by positron emission. The only exception is ^{88}Zr, which decays by electron capture.

Five isotopes of zirconium also exist as metastable isomers: ^{83m}Zr, ^{85m}Zr, ^{89m}Zr, ^{90m1}Zr, ^{90m2}Zr and ^{91m}Zr. Of these, ^{90m2}Zr has the shortest half-life at 131 nanoseconds. ^{89m}Zr is the longest lived with a half-life of 4.161 minutes.

Occurrence

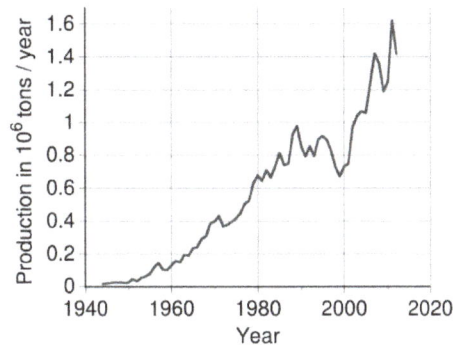

World production trend of zirconium mineral concentrates

Zirconium has a concentration of about 130 mg/kg within the Earth's crust and about 0.026 µg/L in sea water. It is not found in nature as a native metal, reflecting its intrinsic instability with respect to water. The principal commercial source of zirconium is zircon ($ZrSiO_4$), a silicate mineral, which is found primarily in Australia, Brazil, India, Russia, South Africa and the United States, as well as in smaller deposits around the world. As of 2013, two-thirds of zircon mining occurs in Australia and South Africa. Zircon resources exceed 60 million tonnes worldwide and annual worldwide zirconium production is approximately 900,000 tonnes. Zirconium also occurs in more than 140 other minerals, including the commercially useful ores baddeleyite and kosnarite.

Zirconium is relatively abundant in S-type stars, and it has been detected in the sun and in meteorites. Lunar rock samples brought back from several Apollo missions to the moon have a high zirconium oxide content relative to terrestrial rocks.

Production

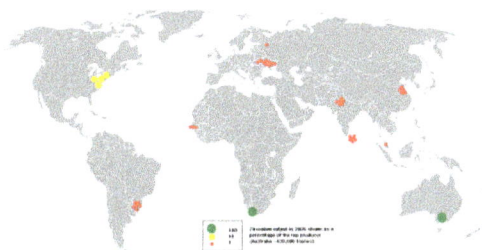

Zirconium output in 2005

Zirconium is a by-product of the mining and processing of the titanium minerals ilmenite and rutile, as well as tin mining. From 2003 to 2007, while prices for the mineral zircon steadily increased from $360 to $840 per tonne, the price for unwrought zirconium metal decreased from $39,900 to $22,700 per ton. Zirconium metal is much higher priced than zircon because the reduction processes are expensive.

Collected from coastal waters, zircon-bearing sand is purified by spiral concentrators to remove lighter materials, which are then returned to the water because they are natural components of beach sand. Using magnetic separation, the titanium ores ilmenite and rutile are removed.

Most zircon is used directly in commercial applications, but a small percentage is converted to the metal. Most Zr metal is produced by the reduction of the zirconium(IV) chloride with magnesium metal in the Kroll process. Commercial-quality zirconium for most uses still has a content of 1% to 3% hafnium. This contaminant is important only in nuclear applications. The resulting metal is sintered until sufficiently ductile for metalworking.

Separation of Zirconium and Hafnium

Commercial zirconium metal typically contains 1–2.5% of hafnium, which is not problematic because the chemical properties of hafnium and zirconium are very similar. Their neutron-absorbing properties differ strongly, however, necessitating the separation of hafnium from zirconium for nuclear reactors. Several separation schemes are in use. The liquid-liquid extraction of the thiocyanate-oxide derivatives exploits the fact that the hafnium derivative is slightly more soluble in methyl isobutyl ketone than in water. This method is used mainly in United States.

Zr and Hf can also be separated by fractional crystallization of potassium hexafluorozirconate (K_2ZrF_6), which is less soluble in water than the analogous hafnium derivative.

Fractional distillation of the tetrachlorides, also called extractive distillation, is used primarily in Europe.

The product of a quadruple VAM (vacuum arc melting) process, combined with hot extruding and different rolling applications is cured using high-pressure, high-temperature gas autoclaving. This produces reactor-grade zirconium that is about 10 times more expensive than the hafnium-contaminated commercial grade.

Hafnium must be removed from zirconium for nuclear applications because hafnium has a neutron absorption cross-section 600 times greater than zirconium. The separated hafnium can be used for reactor control rods.

Compounds

Like other transition metals, zirconium forms a wide range of inorganic compounds and coordination complexes. In general, these compounds are colourless diamagnetic solids wherein zirconium has the oxidation state +4. Far fewer Zr(III) compounds are known, and Zr(II) is very rare.

Oxides, Nitrides and Carbides

The most common oxide is zirconium dioxide, ZrO_2, also known as *zirconia*. This colourless solid has exceptional fracture toughness and chemical resistance, especially in its cubic form. These properties make zirconia useful as a thermal barrier coating, although it is also a common diamond substitute. Zirconium monoxide, ZrO, is also known and S-type stars are recognised by detection of its emission lines in the visual spectrum.

Zirconium tungstate has the unusual property of shrinking in all dimensions when heated, whereas most other substances expand when heated. Zirconyl chloride is a rare water-soluble zirconium complex with the relatively complicated formula $[Zr_4(OH)_{12}(H_2O)_{16}]Cl_8$.

Zirconium carbide and zirconium nitride are refractory solids. The carbide is used for drilling tools and cutting edges. Zirconium hydride phases are also known.

Halides and Pseudohalides

All four common halides are known, ZrF_4, $ZrCl_4$, $ZrBr_4$, ZrI_4. All have polymeric structures and are far less volatile than the corresponding monomeric titanium tetrahalides. All tend to hydrolyse to give the so-called oxyhalides and dioxides.

The corresponding tetraalkoxides are also known. Unlike the halides, the alkoxides dissolve in nonpolar solvents. Dihydrogen hexafluorozirconate is used in the metal finishing industry as an etching agent to promote paint adhesion.

Organic Derivatives

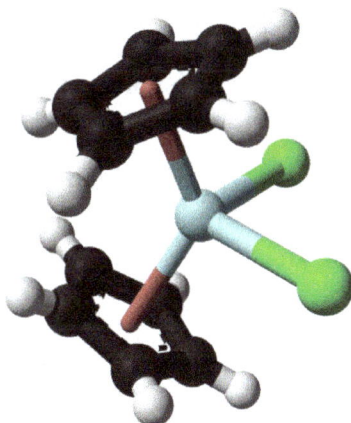

Zirconocene dichloride, a representative organozirconium compound.

Organozirconium chemistry is the study of compounds containing a carbon-zirconium bond. The first such compound was zirconocene dibromide (($C_5H_5)_2ZrBr_2$), reported in 1952 by Birmingham and Wilkinson. Schwartz's reagent, prepared in 1970 by P. C. Wailes and H. Weigold, is a metallocene used in organic synthesis for transformations of alkenes and alkynes.

Zirconium is also a component of some Ziegler-Natta catalysts, used to produce polypropylene. This application exploits the ability of zirconium to reversibly form bonds to carbon. Most complexes of Zr(II) are derivatives of zirconocene, one example being $(C_5Me_5)_2Zr(CO)_2$.

History

The zirconium-containing mineral zircon and related minerals (jargoon, hyacinth, jacinth, ligure) were mentioned in biblical writings. The mineral was not known to contain a new element until 1789, when Klaproth analyzed a jargoon from the island of Ceylon (now Sri Lanka). He named the new element Zirkonerde (zirconia). Humphry Davy attempted to isolate this new element in 1808 through electrolysis, but failed. Zirconium metal was first obtained in an impure form in 1824 by Berzelius by heating a mixture of potassium and potassium zirconium fluoride in an iron tube.

The *crystal bar process* (also known as the *Iodide Process*), discovered by Anton Eduard van Arkel and Jan Hendrik de Boer in 1925, was the first industrial process for the commercial production of metallic zirconium. It involves the formation and subsequent thermal decomposition of zirconium tetraiodide, and was superseded in 1945 by the much cheaper Kroll process developed by William Justin Kroll, in which zirconium tetrachloride is reduced by magnesium:

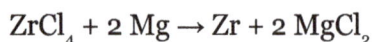

$$ZrCl_4 + 2\,Mg \rightarrow Zr + 2\,MgCl_2$$

Applications

Approximately 900,000 tonnes of Zr ores were mined in 1995, mostly as zircon.

Compounds

Most zircon is used directly in high temperature applications. This material is refractory, hard, and resistant to chemical attack. Because of these properties, zircon finds many applications, few of which are highly publicized. Its main use is as an opacifier, conferring a white, opaque appearance to ceramic materials. Because of its chemical resistance, zircon is also used in aggressive environments, such as moulds for molten metals.

Zirconium dioxide (ZrO_2) is used in laboratory crucibles, in metallurgical furnaces, and as a refractory material. Because it is mechanically strong and flexible, it can be sintered into ceramic knives and other blades. Zircon ($ZrSiO_4$) and the cubic zirconia (ZrO_2) are cut into gemstones for use in jewelry.

Zirconia is a component in some abrasives, such as grinding wheels and sandpaper.

Metal

A small fraction of the zircon is converted to the metal, which finds various niche applications. Because of zirconium's excellent resistance to corrosion, it is often used as an alloying agent in materials that are exposed to aggressive environments, such as surgical appliances, light filaments, and watch cases. The high reactivity of zirconium with oxygen at high temperatures is exploited in some specialised applications such as explosive primers and as getters in vacuum tubes. The same property is (probably) the purpose of including Zr nano-particles as pyrophoric material in explosive weapons such as the BLU-97/B Combined Effects Bomb.

Nuclear Applications

Cladding for nuclear reactor fuels consumes about 1% of the zirconium supply, mainly in the form of zircaloys. The desired properties of these alloys are a low neutron-capture cross-section and resistance to corrosion under normal service conditions. Efficient methods for removing the hafnium impurities were developed to serve this purpose.

One disadvantage of zirconium alloys is that zirconium reacts with water at high temperatures, producing hydrogen gas and accelerated degradation of the fuel rod cladding:

$$Zr + 2\,H_2O \rightarrow ZrO_2 + 2\,H_2$$

This exothermic reaction is very slow below 100 °C, but at temperature above 900 °C the reaction is rapid. Most metals undergo similar reactions. The redox reaction is relevant to the instability of fuel assemblies at high temperatures. This reaction was responsible for a small hydrogen explosion first observed inside the reactor building of Three Mile Island nuclear power plant in 1979, but at that time, the containment building was not damaged. The same reaction occurred in the reactors 1, 2 and 3 of the Fukushima I Nuclear Power Plant (Japan) after the reactor cooling was interrupted by the earthquake and tsunami disaster of March 11, 2011 leading to the Fukushima I nuclear accidents. After venting the hydrogen in the maintenance hall of those three reactors, the mixture of hydrogen with atmospheric oxygen exploded, severely damaging the installations and at least one of the containment buildings. To avoid explosion, the direct venting of hydrogen to the open atmosphere would have been a preferred design option. Now, to prevent the risk of explosion in many pressurized water reactor (PWR) containment buildings, a catalyst-based recombinator is installed that converts hydrogen and oxygen into water at room temperature before the hazard arises.

Space and Aeronautic Industries

Materials fabricated from zirconium metal and ZrO_2 are used in space vehicles where resistance to heat is needed.

High temperature parts such as combustors, blades, and vanes in jet engines and stationary gas turbines are increasingly being protected by thin ceramic layers, usually composed of a mixture of zirconia and yttria.

Positron Emission Tomography Cameras

The isotope [89]Zr has been applied to the tracking and quantification of molecular antibodies with positron emission tomography (PET) cameras (a method called "immuno-PET"). Immuno-PET has reached a maturity of technical development and is now entering the phase of wide-scale clinical applications. Until recently, radiolabeling with [89]Zr was a complicated procedure requiring multiple steps. In 2001–2003 an improved multistep procedure was developed using a succinylated derivative of desferrioxamine B (N-sucDf) as a bifunctional chelate, and a better way of binding [89]Zr to mAbs was reported in 2009. The new method is fast, consists of only two steps, and uses two widely available ingredients: [89]Zr and the appropriate chelate.

Biomedical Applications

Zirconium-bearing compounds are used in many biomedical applications, including dental implants, knee and hip replacements, middle-ear ossicular chain reconstruction, and other restorative and prosthetic devices.

Zirconium binds urea, a property that has been utilized extensively to the benefit of patients with chronic kidney disease. For example, zirconium is a primary component of the sorbent column dependent dialysate regeneration and recirculation system known as the REDY system, which was first introduced in 1973. More than 2,000,000 dialysis treatments have been performed using the sorbent column in the REDY system. Although the REDY system was superseded in the 1990s by less expensive alternatives, new sorbent-based dialysis systems are being evaluated and approved by the U.S. Food and Drug Administration (FDA). Renal Solutions developed the DIALISORB technology, a portable, low water dialysis system. Also, developmental versions of a Wearable Artificial Kidney have incorporated sorbent-based technologies.

Zirconium cyclosilicate is under investigation for oral therapy in the treatment of hyperkalemia. It is a highly selective oral sorbent designed specifically to trap potassium ions in preference to other ions throughout the gastrointestinal tract.

Defunct Applications

Zirconium carbonate ($3ZrO_2 \cdot CO_2 \cdot H_2O$) was used in lotions to treat poison ivy but was discontinued because it occasionally caused skin reactions.

Safety

Although zirconium has no known biological role, the human body contains, on average, 250 milligrams of zirconium, and daily intake is approximately 4.15 milligrams (3.5 milligrams from food and 0.65 milligrams from water), depending on dietary habits. Zirconium is widely distributed in nature and is found in all biological systems, for example: 2.86 µg/g in whole wheat, 3.09 µg/g in brown rice, 0.55 µg/g in spinach, 1.23 µg/g in eggs, and 0.86 µg/g in ground beef. Further, zirconium is commonly used in commercial products (e.g. deodorant sticks, aerosol antiperspirants) and also in water purification (e.g. control of phosphorus pollution, bacteria- and pyrogen-contaminated water).

Short-term exposure to zirconium powder can cause irritation, but only contact with the eyes requires medical attention. Persistent exposure to zirconium tetrachloride results in increased mortality in rats and guinea pigs and a decrease of blood hemoglobin and red blood cells in dogs. However, in a study of 20 rats given a standard diet containing ~4% zirconium oxide, there were no adverse effects on growth rate, blood and urine parameters, or mortality. The U.S. Occupational Safety and Health Administration (OSHA) legal limit (permissible exposure limit) for zirconium exposure is 5 mg/m³ over an 8-hour workday. The National Institute for Occupational Safety and Health (NIOSH) recommended exposure limit (REL) is 5 mg/m³ over an 8-hour workday and a short term limit of 10 mg/m³. At levels of 25 mg/m³, zirconium is immediately dangerous to life and health. However, zirconium is not considered an industrial health hazard. Furthermore, reports of zirconium-related adverse reactions are rare and, in general, rigorous cause-and-effect relationships have not been established. No evidence has been validated that zirconium is carcinogenic or genotoxic.

Among the numerous radioactive isotopes of zirconium, ^{93}Zr is among the most common. It is released as a product of ^{235}U, mainly in nuclear plants and during nuclear weapons tests in the 1950s and 1960s. It has a very long half-life (1.53 million years), its decay emits only low energy radiations, and it is not considered as highly hazardous.

References

- Lide, D. R. (2000). "Magnetic susceptibility of the elements and inorganic compounds" (PDF). CRC Handbook of Chemistry and Physics (81st ed.). CRC Press. ISBN 0849304814.

- Dickin, A. P. (2005). "In situ Cosmogenic Isotopes". Radiogenic Isotope Geology. Cambridge University Press. ISBN 978-0-521-53017-0.

- Greenwood, Norman N.; Earnshaw, Alan (1997). Chemistry of the Elements (2nd ed.). Butterworth-Heinemann. p. 217. ISBN 0-08-037941-9.

- Green, John A. S. (2007). Aluminum Recycling and Processing for Energy Conservation and Sustainability. ASM International. p. 197. ISBN 1615030573.

- Polmear, I. J. (2006). "Production of Aluminium". Light Alloys from Traditional Alloys to Nanocrystals. Elsevier/Butterworth-Heinemann. pp. 15–16. ISBN 978-0-7506-6371-7.

- "Elements of Chemical Philosophy By Sir Humphry Davy". Quarterly Review. John Murray. VIII: 72. 1812. ISBN 0-217-88947-6. Retrieved 10 December 2009

- Krebs, Robert E. (2006). The History and Use of Our Earth's Chemical Elements: A Reference Guide (2nd ed.). Westport, CT: Greenwood Press. ISBN 0-313-33438-2.

- Matthew J. Donachie, Jr. (1988). Titanium: A Technical Guide. Metals Park, OH: ASM International. Appendix J, Table J.2. ISBN 0-87170-309-2.

- Greenwood, Norman N.; Earnshaw, Alan (1997). Chemistry of the Elements (2nd ed.). Butterworth-Heinemann. p. 962. ISBN 0-08-037941-9.

- Seong, S.; et al. (2009). Titanium: industrial base, price trends, and technology initiatives. Rand Corporation. p. 10. ISBN 0-8330-4575-X.

- Coates, Robert M.; Paquette, Leo A. (2000). Handbook of Reagents for Organic Synthesis. John Wiley and Sons. p. 93. ISBN 0-470-85625-4.

- Hartwig, J. F. (2010) Organotransition Metal Chemistry, from Bonding to Catalysis. University Science Books: New York. ISBN 189138953X

- Matthew J. Donachie, Jr. (1988). TITANIUM: A Technical Guide. Metals Park, OH: ASM International. pp. 16, Appendix J. ISBN 0-87170-309-2.

- Moiseyev, Valentin N. (2006). Titanium Alloys: Russian Aircraft and Aerospace Applications. Taylor and Francis, LLC. p. 196. ISBN 978-0-8493-3273-9.

- Kleefisch, E.W., ed. (1981). Industrial Application of Titanium and Zirconium. West Conshohocken, PA: ASTM International. ISBN 0-8031-0745-5.

- Gruntman, Mike. Blazing the Trail: The Early History of Spacecraft and Rocketry. Reston, VA: American Institute of Aeronautics and Astronautics. p. 457. ISBN 1-56347-705-X.

- Cotell, Catherine Mary; Sprague, J. A.; Smidt, F. A. (1994). ASM Handbook: Surface Engineering (10th ed.). ASM International. p. 836. ISBN 0-87170-384-X.

- Solomon, Robert E. (2002). Fire and Life Safety Inspection Manual. National Fire Prevention Association (8th ed.). Jones & Bartlett Publishers. p. 45. ISBN 0-87765-472-7.

- Hammond, C. R. "Elements" in Lide, D. R., ed. (2005). CRC Handbook of Chemistry and Physics (86th ed.). Boca Raton (FL): CRC Press. ISBN 0-8493-0486-5.

- Mining, Society for Metallurgy, Exploration (U.S) (5 March 2006). "Distribution of major deposits". Industrial minerals & rocks: commodities, markets, and uses. pp. 265–269. ISBN 978-0-87335-233-8.

- Storer, Frank Humphreys (1864). First Outlines of a Dictionary of Solubilities of Chemical Substances. Cambridge. pp. 278–80. ISBN 978-1-176-62256-2.

Permissions

Index